GOD IS:

A MATTER OF FACT

THE SCIENTIFIC THEORY OF

SUPREME INTELLIGENCE

BY

MARC WATSON, P.E.

A STUDY OF THE

"LAWS OF NATURE AND NATURE'S GOD"*

*FROM
THE DECLARATION OF INDEPENDENCE
OF
THE UNITED STATES

afflatus

This Book is Dedicated to

Sherre

My Wonderful, Compassionate Wife

My Best Friend

The Champion of my Life.

"I want to know how God created this world. I'm not interested in this or that phenomenon, in the spectrum of this or that element. I want to know His thoughts, the rest are details."

- Albert Einstein. From E. Salaman, "A Talk with Einstein," *The Listener* 54 (1955): 370-3

TABLE OF CONTENTS

PREFACE

I am an Engineer. So I must begin by begging your forgiveness, for I am naturally wired to be boring (just ask my wife). If you are continuing to read, then I will assume that you have accepted my apology in advance of tedium and are ready to move ahead.

I believe that anytime someone takes the time to read a work such as this, they are entitled to know a little about the author and the journey taken in placing these words on this paper (or in this E-Book). Therefore, at the risk of immediately losing the reader, here is the short version on how *GOD IS: A Matter of Fact*, came to be (as short as an engineer can make it!)

After receiving a degree in Ocean Engineering from the United States Naval Academy, I spent a number of years in practice and passed two national level examinations to obtain my registration as a Professional Engineer. At that point, I thought I knew everything. What I found out, in fact, was that I was just beginning to learn anything. Isn't it funny that we are so amazingly smart when we are young and seem to grow in stupidity as we grow in years?

I had the good fortune to work as an engineer for two very large organizations; the U. S. Navy and Sverdrup Corporation (now Jacobs Engineering), along with a highly rewarding opportunity to own a small engineering company. Yet my real training in engineering came after practicing for nearly twenty years, when I started working with the entertainment giant Universal Studios.

You see, up until that point, I had viewed engineering as a matter of the structures or systems being designed or developed. Engineering was about matter; the stuff things are made of, and the forces governing that matter. However, in my first design experience with Universal Studios, I was humbled to learn that matter wasn't all that mattered. What in fact mattered more than matter was the

Human Experience. Everything we were developing was a full contact sport with Human Interaction.

I started on something simple, Jurassic Park©! Mark Woodbury, now President of Universal Creative, demanded that every element of everything we designed at Universal focus on the guest – what the industry refers to as the Guest Experience. How would people engage with the rides and shows? How would they behave in the queue lines, and how could we make that long and sometimes boring process enjoyable? Guest safety and guest satisfaction drove everything we did; and drove most of us up a wall.

As the director for the Amazing Adventures of Spiderman©, I was privileged to lead a creative team that employed the latest technologies to design what was and still is one of the most advanced ride systems in the world. Yet our best engineering was sliced, diced, and shredded to ribbons by our operations, technical services, and safety teams. They had to deal with a much more sophisticated and a seemingly less predictable force: Human Intelligence.

For years I have continually studied and practiced engineering confident in my understanding of the Natural Laws of the Universe – the Fundamental Interactions of Matter. Those Fundamental Interactions initiate the forces that make all of the systems we design work. Gravity, electricity, mechanical forces, chemical reactions; all flow from the Fundamental Interactions. They were the forces that drove our world – they were the forces that mattered.

My experience with Universal Studios opened my eyes to an entirely different set of forces; the forces that drive Human Behavior. These forces are just as real as the natural laws of matter – and just as natural. Yet these forces are different. I found that if we analyzed human behavior we could indeed evaluate these forces and how they affect human interaction with material systems. We could actually predict when people would have fun (as well as when they might get aggravated).

This concept of engineering for Human Interactions and Human Behavior is now called Human Factors Engineering (pretty original). This discipline is relatively new, having been formally introduced with the creation of The Institute of Ergonomics and Human Factors (formerly the Ergonomics Society) in 1949. The discipline is becoming increasingly more important as we recognize that we should focus more effort on designing systems around Human Behavior; instead of simply trying to constrain Human Behavior to conform to our systems.

My engineering career has taken another twist over the past several years as I have become more involved in the design of learning systems, particularly immersive learning experiences for Science, Technology, Engineering, and Math ("STEM"). This work has resulted in a more rigorous understanding of Human Behavior in the Learning process (versus the "Educating" process). My experience in this journey will be the grist for another day. I bring it up only to illustrate the full circle of my personal journey in Science. I started my travels in wonderment in junior high school science classes. I am now working to inspire and engage young students in the same level of excitement about the fascinating Universe we live in and the gift of Life we have all been granted.

My experiences and observations have led me to the conclusion that Intelligence is a Fundamental Phenomenon of Nature. This is the core Scientific Theory of this Book. While this conclusion may seem obvious to many, it is generally rejected as scientific, and is therefore disallowed in public education.

The Theory of Supreme Intelligence recognizes that Fundamental Intelligence is a fifth Fundamental Phenomenon that is independent of the four Fundamental Interactions of Matter. There are forces that flow from Supreme Intelligence that cannot be explained by the Laws of Matter; yet they are part of the Laws of Nature. People are more than just matter. Human Intelligence is more than merely a response to four Fundamental Interactions. The more we understand the Phenomenon of Supreme Intelligence and the forces that drive Human Behavior, the better equipped we are to design the systems that deal with Human Interaction.

I now understand that God did not design the Universe for the sake of matter; but for the sake of Intelligent Beings. God Created the Universe and Life with one specific thing in mind: The Guest Experience.

Thank you. I want to say a special thank you to four people who, in addition to my wife Sherre, helped me get this manuscript out of my computer.

Connie Byrd is one of the smartest math teachers I have ever known. Her educational credentials and position as the head of a high school math department are testament to her knowledge on the subject. However, her true gift lies in her ability to take complex mathematical logic and distill it into understandable knowledge for her students. She has helped me find logic in what at first glance appears illogical.

Barbara Angle is a chemical engineer, and is steeped in the knowledge of Fundamental Interactions and the complexities of the chemical reactions that drive much of Nature and Life. In her support of this work, she delicately posed inquiry with regard to points raised (I have a frail ego), without simply stating the obvious: Are you crazy?

Sandy Sprott and Greg Watson served as a final sounding board, assisting me in finalizing the overall presentation of this work. Sandy is a creative genius. Her mind simply sees things that others miss. Greg Watson is a practical engineer – things have to make sense.

Thank you Connie, Barb, Sandy, and Greg.

Finally, and most importantly, thank you to you, the reader. I hope that this work provides a Guest Experience for you such that you will not regret the time you invested reading it.

Thank you, and God Bless you.

Marc Watson.

Suggestions on Using This Book

There are a few thoughts I adopted when writing this book that you may wish to consider when reading or using it.

1. **It was written to be read.** This sounds obvious, but I run across a number of people who will make assumptions about or begin debating a book based on the title, a back cover description, or the Table of Contents. This is not meant to be a coffee table book. I strongly encourage discussion, debate, and even critical analysis of what is presented herein. However, such analysis should take place after the content herein has been read.

2. **Feel free to Explore.** While I have attempted to organize this engineering treatise as a methodical presentation of the Theory, this approach can become plodding at times. Therefore, you may find this work equally as valuable by triggering interest, discussion, or critical thinking in the theories and concepts introduced in a specific chapter or section. If there is a topic that is of particular interest, start your exploration there. That approach may leave some gaps in the logical flow of the presentation, but if it promotes engagement in the subject matter, go ahead. Once you are engaged, you may choose to go back to those skipped sections to fill in the gaps.

3. **It should be approached with an open mind.** Scientific Method relies on an open mind and a willingness to question, consider, debate, and continuously learn. I certainly don't expect everyone to agree with all that is presented. In fact, dissenting opinion is welcomed and is part of a strong scientific study. Likewise, I would request that the information presented not be summarily

dismissed. The axioms, theories, postulates, hypotheses, assumptions, and conditions are presented so that they can be discussed.

4. **It is intended for further Learning.** I have Capitalized a number of terms in this book, not because I forgot the basics of punctuation, but because I wanted to emphasize certain terms that create interesting areas for additional learning. Also, some of these terms take on the significance of proper nouns in the context they are used, and therefore warrant capitalization. Several reviewers suggested that such capitalization detracted from the reading. Hopefully, the capitalized emphasis for learning will offset any distraction in reading. This book is intended to promote further Learning!

5. **Tradition, Reason, Experimentation.** You will see that I use these three terms throughout the book to indicate one of three Learning Channels – or sources of the knowledge we acquire over time. Think of it as watching three different news channels on television. If you watch news channels offering three different perspectives, you will acquire Balanced Knowledge on a subject. All three channels consider their news reliable. However, since a single channel flows from a particular perspective, it is very difficult for that news to be balanced. Watching a single channel for all your news will skew your perspective. The same is true for Learning. This issue will be dealt with in detail in Part Two of this book. These terms will also be capitalized so you will know which "Learning Channel" I am referring to.

With these suggestions in mind:

LET THE SCIENTIFIC STUDY BEGIN

PART ONE: GOD IS

Chapter 1 - GOD IS

THE STATEMENT

Why do we have so many issues with such a simple statement: "GOD IS"? Merely expressing that GOD IS evokes wide-ranging thoughts and emotions. Most people intuitively know that GOD IS. However, for many the discussion or study of God's existence is typically isolated to a place of worship.

For others the mere assertion that GOD IS causes them to see red. There are some who simply cannot or will not accept any concept of a Supreme Intelligent Being that created the Universe and Life. Others disavow God because of questions regarding the historical accuracy of some religious traditions. For many of them, GOD IS a matter of fantasy or fallacy; not fact.

The question of Who God Is and What God Is has enthralled human thought since humans began thinking. So has the debate regarding Whether God Is. Yet this wonderment about God is less about God than it is about us. We wonder where we came from; why we are here; and where we are going. Is there a purpose to the Universe or Life? Our quest for answers about God is really a desire to find out more about ourselves. This book is not intended to answer these questions or settle the debate. To the contrary, it is intended to encourage the questions and stimulate debate by reintroducing the Statement into the public discourse as a matter of scientific study.

This work will demonstrate through unequivocal evidence that God is not a matter of fantasy or superstition; not a philosophical construct or religious legend. Certainly not a figment of imagination. This study will demonstrate that **GOD IS: A Matter of Fact**.

THE QUESTION

The Question, which is actually an entire series of questions, has been around since the beginning of time. Where do we come from? Why are we here? What created the Universe? More importantly, what causes the Behavior of Life? How can we improve Human Interactions and be better stewards of Planet Earth?

The theories are as old as the Question. Our desire for answers is not merely an exercise in mental gymnastics. As we learn more about the beginning of the Universe and the origin of Life, we uncover practical knowledge that allows us to discover new solutions to problems; address challenges that cause suffering and strife; and advance the Human Condition.

During the previous several centuries we have vastly improved our knowledge of the Universe and Life, the two realities that are central to the Question. Yet, with as much as we know, there is still so much we don't know. While we have made significant strides in our understanding of both the macro-Universe and the micro-world, we have made little progress in advancing our understanding of Human Interaction.

One reason for this lack of advancement in our understanding of Human Interaction is that in the last one hundred years scientific study of the Question has been restricted to theories rendering man as the supreme intelligence in the Universe. Currently most public scientific study of the Question will not allow for any theory regarding the existence of any intelligent force superior to human intelligence. The only scientific theories allowed dictate that all answers to the Question be derived from the unintelligent interactions of Matter. Such theories require that Intelligence be derived from Matter, without even considering that Matter could be derived from Intelligence.

As we continue to search for answers to the Question, we must use all of our reason, observation, and experimentation to explore all reasonable theories. We must allow the full range and power of intelligence to deal with our challenges. However, in seeking answers through the application of intelligence, we are faced with one of the most perplexing of all Questions: "Is there an Intelligence anywhere in the Universe that is superior to man's?"

THE THEORY

The Phenomena of the Universe and Life are clearly observable; both are Matters of Fact. Our ability to address the Question posed above is to a great extent dependent upon our understanding of the origin of these two Phenomena. In order to further our understanding of the Universe and Life, we shall draw upon another equally observable Phenomenon, Intelligence. This Engineering Treatise recognizes the Phenomenon of Intelligence; and evaluates Intelligence as a Fundamental Phenomenon of Nature and therefore a plausible Source of the Fundamental Forces and Interactions that created and influenced the development of the Universe and Life.

As with any scientific study, we shall introduce a theory based on initial reasoning with regard to the Statement and Question. We shall then apply our best current knowledge and observations on the subject to advance our learning, leading to Balanced Knowledge (more on that term shortly). We will step through a series of Axioms that will influence our knowledge on the Theory, either positively or negatively. As the study advances, we will present a Postulate that starts to coalesce all of our reasoning into a supportable scientific argument.

Therefore, we will begin this scientific study by offering the following theory.

Theory 1.1: The Theory of Supreme Intelligence: A Singular Supreme Intelligent Phenomenon Initiated the Universe and Originated Life.

WHY ASK?

"Why Ask?" you ask (Or if you didn't, I'll ask)? This Theory doesn't seem to present anything new. The concept has been around for centuries. Many people would accept the Theory as a fact and simply move on. Why do we need to ask the Question or provide a Theory for the existence of God? After all, GOD IS! Why do we need this book? The answer lies in Recent History and Present Direction.

Recent History. One hundred years ago the World was in Serious Trouble! Decades earlier, Friedrich Nietzsche declared "God is Dead" and Karl Marx asserted that real happiness demanded the abolition of religion. By the turn of the 20th century, these men's "godless" social and economic science had inspired generations of revolutionary thought in a restless world; particularly in Central Europe and Asia. At the same time, advances in physics and chemistry demonstrated the power of science in addressing the challenges to the Human Condition. God was rapidly being displaced by scientists, many espousing the religion of Atheism. These titans of science became the masters of all Reliable Knowledge in a world dominated by the Fundamental Interactions of Matter.

Science had pushed God to the sideline and the result was catastrophic. One third of the world was plunged into oppressive anti-God Communism. Hitler used Nietzsche's super race theory to villainize an entire culture and ignite the most destructive global war ever. Nuclear arms were introduced to a world of growing cultural intolerance.

Fast forward to the present day and we find the world on the same slippery slope as a hundred years ago. This time, however, the United States is following a similar trajectory that Europe followed at the turn of the last century. Our diverse society is growing increasingly more intolerant of cultural, ethnic, and religious differences. Battle lines are being drawn between Capitalism and Socialism; as politicians and media fan the flames of smoldering class warfare. Science, through the statements of notable scientists, has more aggressively asserted its position as the supreme intelligence of the Universe, rejecting any notion or acknowledgment of any other Supreme Intelligent Being. In America, where God is recognized as the source of Unalienable Rights and the Blessings of Liberty, we can't even consider the scientific evidence of God in high school classrooms.

The primary reasons for this situation can be found in two alarming biases in our **Present Direction;** specifically **Restricted Science** and **Creation Teaching.**

Restricted Science. Over the last century there has been a concerted effort to eliminate any discussion of God or acknowledgement of the existence of God from scientific study regarding our Universe and the forces that drive Nature and Life. The desire to remove God from public scientific study is driven by a portion of the scientific community that will not allow a theory such as The Theory of Supreme Intelligence to enter the scientific discourse regarding the beginning of the Universe or the creation of Life. Highly educated, articulate, and charismatic scientists have become spokesmen for a single sided debate limited to their point of view. They have restrained scientific study to their belief system and experience. We shall refer to such restrained study herein as Restricted Science.

Restricted Science could flow from the fact that many who oppose the scientific study of concepts such as Supreme Intelligence profess the religion of Atheism, a belief system that openly denies the existence of God. Their belief system carries over to their science, and they are therefore unwilling to allow any such consideration in scientific study. Or possibly they simply fear such a debate may expose weaknesses in their own proposed theories or beliefs; or possibly even disprove them.

No matter the motive, limiting scientific discussion for any reason is Bad Science. Therefore, since the Theory of Supreme Intelligence has been proffered, we should let the scientific study commence!

Creation Teaching. One of the most naive restrictions against the scientific study of concepts such as Supreme Intelligence and God in public schools is that we don't "Teach Creation" in the public classroom. That is simply false. We do in fact teach Creation in every public middle school and high school in the United States. It is deeply embedded in our science curriculum. The confusion lies in three little letters, "…ism": as in Creationism. While Creationism is disallowed, Creation is certainly being taught.

That bears repeating: We teach Creation in public schools! The problem is that we teach Creation from one singular belief system – the complete denial of any superior intelligence in the Universe.

We currently teach about the Origin of the Universe and the

Creation and Evolution of Life in both introductory and advanced science classes. In fact, two scientific theories emerging in the last half century focus at the heart of the Question and address the very beginning of the Universe. They are the Grand Unified Theory (GUT) and the Theory of Everything (TOE). No matter what you call such theories, these theories by any other name are still Creation Theories.

Unfortunately, as stated above, Restricted Science is at work in the teaching of Creation. We have restrained the study of Creation concepts to only those theories involving inanimate, unintelligent interactions as the cause for all the forces and behaviors in the Universe. Such restraint is inconsistent with the evidence and even inconsistent within some of the basic theories we now teach.

This problem is clearly illuminated in the current scientific study of Evolution in public classrooms throughout the United States. In *The Origin of Species*, Charles Darwin's seminal work on Natural Selection and Evolution, his observations regarding the power of "Man's Selection" (further defined herein as Intelligent Selection) form the foundation for his Theory of Natural Selection. Darwin reflects on the wide variety of plant and animal species created by Man's Selection in the Introduction of that work, and dedicates the entire first chapter of his book to this powerful form of Selection. Yet this very important underlying principle for the Theory of Natural Selection in Evolution is not even introduced in the public classroom discussion of Evolution. Honestly, have you ever heard of Darwin's concept of "Man's Selection" with regard to the Evolution of Species or variants in plants and animals? (We will delve more deeply into this discussion in Chapter 15).

How can we encourage the rigorous scientific study of the principles of Evolution without even allowing for the discussion of the foundational observation leading to the Theory of Evolution? Our public allowance for scientific study of all matters relating to the Creation of the Universe and Life has been and continues to be limited to a singular school of thought: It was all a random spontaneous accident!

It is particularly ironic that in the United States we inappropriately apply a First Amendment prohibition on religion to restrict our scientific study of Creation. We have prevented the introduction or critical analysis of theories that would demonstrate the existence of a Supreme Intelligent Phenomenon. Such theories have substantial scientific merit, and offer the opportunity to further expand our knowledge of not only Creation, but also the complexities and challenges of Human Interaction.

AN ENGINEERING TREATISE

This book takes a scientific approach to evaluating the Statement "GOD IS: A Matter of Fact", the Question, and the Theory.

GOD IS a matter of fact; and therefore not only withstands but clearly emerges from rigorous scientific study. Similarly, Human Behavior is not simply a response to the Fundamental Interactions of Matter. Humans are Intelligent Beings with the Free Will to use our intellect for the greater good; or to do unspeakable evil.

We have already discussed the Question and proposed a reasonable Theory for evaluation. We will apply Scientific Method through analysis, logic, observation, and critical thinking to arrive at Axioms that allow us to advance learning with regard to the Theory. Alternate theories and counter arguments will be presented in an attempt to balance the analysis. We will let the evidence take us wherever it takes us.

This book does not rely on any religious dogma or practices, or refer to any of the stories, history, or revelation that provide the framework for religious traditions. There will be no discussion of miracles or prophecies.

This is an Engineering Treatise. Engineers apply Balanced Knowledge to address problems and find solutions to the wants and needs of individuals and societies; producing products, processes, designs, and services. Designed systems not only include machines and structures, but also the internal and external forces interacting with those machines and structures. These designs can include social systems, such as Economic and Political systems; as well as the more traditional structures and mechanical or electrical systems we typically consider as engineering.

Engineers deal predominantly with **Forces** and **Interactions**, two terms that you will find used liberally throughout this work. We are certainly concerned with and will discuss Matter, Mass, and Energy; the focus of much of the current scientific study on the Question. However, Forces drive Behavior; and therefore our observations and analysis will focus on the Behavior of the systems we are evaluating.

Behavior is another key term you will encounter frequently. All of Life, in fact the entire Universe, exhibits Behavior. Behavior is what we observe. Behaviors are the actions of anything that take place over time. Individuals demonstrate Behavior, as do entire Systems; from Solar Systems to galaxies; from atomic structures to complex molecules. Earth demonstrates Behavior; shifting tectonic plates, ocean tides, jet streams and currents, to name a few. Most importantly, Humans demonstrate Behavior. Improving the Human Condition requires an understanding of the Forces that drive Human Behavior.

Forcing Behavior. Forces drive behaviors. Engineers must recognize and design for the wide variety of forces that continuously drive behavior. The behavior may be that of a large suspension bridge, a space vehicle on a mission to a far away planet, or tourists in a queue line at a theme park. When engineers design systems, the behavior of the system and the things that interact with the system are what we must consider.

Most of the forces we encounter are driven by the four Fundamental Interactions of Matter; including Gravity, Strong, Weak, and Electromagnetic. These Fundamental Interactions seem to answer most of the questions regarding Matter, Mass, and Energy. However, observing the Universe and Life demonstrates behaviors that simply cannot be attributed to the four Fundamental Interactions of Matter.

Science sometimes adds large unobserved unknowns, or "plugs", to reconcile these differences. In other words, if an observation indicates a behavior that is inconsistent with what we think we know, we sometimes plug in something we don't know to close the gap between what we have observed and what we currently think we know. This is the reason for the hypothesis of Dark Matter, which will be discussed in more detail a little later.

An equally reasonable alternative to these "plugs" could be the existence of another Fundamental Phenomenon that initiates forces leading to the Behaviors we observe. This is part of the rationale for the scientific study of the Theory of Supreme Intelligence; since, as we will discover, Intelligence can certainly initiate forces that are well beyond the limits of the Fundamental Interactions.

I have attempted to follow an engineering approach to the pragmatic review of definitions, theories, logic, and observations in the context of science and physics. As a work of engineering, we will observe Behaviors of systems, and from those Behaviors determine what Forces are at work. This allows us to improve our understanding of systems in constant motion, and thereby improve the designs of future systems.

I used the term "attempted" purposely, since the reader may feel that my observations, logic, or reasoning fall short in some regard; and are open to debate. Certainly that will be the case, and such debate is welcomed; for it means that you are considering the evidence and continuing the formulation of your own Balanced Knowledge with regard to the theories presented. If you feel there is other logic or reasoning that is more applicable to the analysis herein, then you should engage in the debate.

A Scientific Study for Public Discussion. This Engineering Treatise was written in a manner appropriate for use as a text or reference for a study, lesson plan, or course of instruction in public school science classrooms. Additional study topics are also introduced in this work, promoting further scientific research and discussion on a variety of concepts presented herein; ranging from Interactions, Forces, and Energy; to the Mass-Energy Equivalence and SpaceTime Continuum; to String Theory and Quantum Entanglement.

We must be willing and able to openly and honestly discuss, in the context of a scientific study, the axioms, theories, and postulates presented herein and the existence of God as a Phenomenon of the Universe, our world, and our individual lives. The goal of this book is to catalyze a more public discourse on this subject, a discourse that does not confine itself to specific religious dogma, restricted traditions, or intolerance.

My hope is that this simple work will encourage people to discuss the existence of God in an open, receptive, tolerant, and respectful manner. Only then will we be able to advance our understanding of how God interacts through Natural Laws and Life Forces. Only then will we be able to answer those questions that deal with the Human Condition and our very existence.

A MATTER FOR SCIENCE

The Theory of Supreme Intelligence is a matter for science. We must ensure that studies such as this are part of every high school science curriculum in the United States. Not sociology, not philosophy (although these subjects should also engage the dialogue) - Science! This book will demonstrate that the scientific study of this Theory is warranted; and the resulting Postulate is both reasonable and demonstrable through a methodical review of physics, math, astronomy, and other established science. It is not a book of religious traditions or ancient stories. You will not be treated to lyrical prose or prophetic wonderment (or bewilderment). This is written by an engineer, so be prepared for some pretty boring stuff.

We should not allow Restricted Science to prevent the open and deliberate discussion of all potential theories answering the fundamental questions of the Universe and Life. One of the most dangerous things we can do in science is adopt any singular rule, regulation, or limitation that restricts scientific study or eliminates criticism, reason, experimentation, or advancement.

Falsifiability. Such is the case for a scientific theory called Falsifiability, espoused by a 20th Century Austro-British Philosopher, Sir Karl Popper; and the misapplication of this concept by, of all institutions, U.S. Courts; the guardians of Freedom of Thought. Popper earned a doctorate in psychology and started teaching secondary school; during which time he published his first book, *The Logic of Scientific Discovery,* in 1934. In that book, Popper criticized several of the accepted forms of Scientific Method and introduced his own concept of Falsifiability. This scientific philosophy contends that for a hypothesis to stand the rigor of Scientific Method, it should be testable through observation or experimentation such that if it is false, it can be proven to be false.

Popper was attempting to repudiate reliance upon inductive reasoning as proof of scientific theory. Popper's work in general, and his concept of Falsifiability specifically, gained traction and wider acceptance when, following World War II, he was appointed Professor of Logic and Scientific Method at the University of London.

Fast forward fifty years or so and we find U.S. Courts up through the U.S. Supreme Court using the "Falsifiability" principal as a "relevant factor" under the Daubert Standard in determining whether trial evidence is "Scientific Knowledge". Ironically, Popper himself asserted that "unfalsifiable" statements are also important in science (directly contradicting the Daubert Standard). Sir Karl's own philosophical journey on the subject demonstrated a work in progress. For example, at one point Popper doubted the falsifiability of Natural Selection. However, Popper later recanted, stating that:

> "I have changed my mind about the testability and logical status of the theory of natural selection; and I am glad to have an opportunity to make a recantation".
> - Popper, K (1978). "Natural selection and the emergence of mind". *Dialectica* (32): 339-355

Despite Sir Karl's change of heart, it is difficult to draw up a scientific experiment that can falsify, refute, or test Natural Selection as an evolutionary process. Yet, Evolution is considered science (and I concur that Evolution is a subject for scientific study).

The Falsifiability standard also fails miserably in evaluating other issues for validity of scientific and mathematical study. For example, it would be impossible to set up an experiment to "Falsify" the universality of gravity. Yet universal gravity is accepted as one of the fundamental interactions for all of science.

Thank goodness Sir Karl changed his mind about Darwin's Theory, and that the Supreme Court was never faced with the question of whether gravity is a matter for scientific study.

Generally Accepted Science. An even more common cause of Restricted Science is the "test" that scientific study should only include those subjects that are generally accepted by the community of scientists. This is in fact completely contrary to scientific study. Science is supposed to advance our Balanced Knowledge not only by studying what we think we know, but also by exploring alternatives to current theories based on new observations or reasoning. Science in fact explores concepts that are NOT generally accepted. The initial discoveries of Copernicus, Galileo, Newton, and Einstein were at the edge of or outside the body of Generally Accepted Science.

We have an enlightening example of the restrictive nature of "Generally Accepted Science" in the work of Gabrielle Emilie Le Tonnelier de Breteuil, marquis du Chatelet (she was French). For the next science test, you must memorize her name. This young woman of the early Eighteenth Century had the audacity to challenge Generally Accepted Science in questioning the formula for Kinetic Energy supported by a predominance of the scientific community, including the likes of Newton and Voltaire. This concept is one of the most important principles in physics. At that time, Kinetic Energy was considered to be a function of a body's mass multiplied by its velocity:

$$Ek = mv$$

However, based on the experiments of other scientists, including Gottfried Leibniz and Johann Bernoulli, Emilie du Chatelet argued that the factor was not simply the velocity, but the velocity squared; resulting in an order of magnitude difference in the Kinetic Energy of a body in motion.

$$Ek = \tfrac{1}{2}mv^2$$

This simple change makes all the difference in the world with regard to our understanding of energy, leading to Einstein's further definition of the Mass Energy Equivalence. Despite du Chatelet's extensive knowledge on the subject, including her complete translation of Newton's *Principia Mathematica* into French, her argument regarding Kinetic Energy was largely ignored by Generally Accepted Science for another century. She was posthumously vindicated when she was finally acknowledged to be right!

If we restrict scientific study to Generally Accepted Science, we will prevent one of the most important elements of Scientific Method – **New Discovery.**

THIS BOOK IS NOT ...

This is not a Definitive Work. This book is certainly not "the definitive work" on the subject. In fact, it is written with the opposite purpose in mind. It will not complete discussion on the statement GOD IS; but instead will hopefully reignite public scientific study on the concept that GOD IS. This was written to encourage further learning on the subject. We must bring scientific discussion of GOD out of the proverbial closet.

This is NOT Religion! This is not a work on religion, of religion, or about religion, and does not present any view based on religious dogma, practice, or philosophy. Therefore, it does not fall under any perceived or stated limitation of the First Amendment. It approaches the Statement from an entirely secular scientific perspective, and therefore offers an opportunity for presentation, discussion, and critical thinking on the subject in public schools and for public education.

God is not religion; and religion is not God. God existed long before man, and long before religion. God is not mentioned in the First Amendment, nor is a public scientific study of God restricted by the First Amendment. Indeed, one of the stated objectives of the Constitution is to "secure the Blessings of Liberty"; and at the time of the writing of the Constitution (and even to most of us today), the clear meaning of the term "Blessings" bestowed a very special meaning to the subsequent term, "Liberty"; thus clearly demonstrating that Liberty flowed from a higher authority; God. Liberty is therefore not subject to the whims of government. It was and is an Unalienable Right, flowing from the "Laws of Nature and Nature's God."

Of equal significance in this argument is the fact that every State that ratified the U.S. Constitution and the Bill of Rights included a clear recognition of the existence of God in their own State's Constitution. There is no way one can draw from logic or reason, or otherwise infer that the States that ratified the Constitution had any intention of preventing the recognition of God through the First Amendment prohibition. The First Amendment provides for the tolerance of all religious beliefs or for no religious belief; it does not address the existence of God.

Finally, I think you will find that many religious leaders reject this work as not subscribing to their specific religious traditions with regard to how God is revealed to man, or interacts in our lives (I can almost hear their clamoring now). I believe we could find a long line of religious leaders who would testify that this book is not religion. If it does not walk like a duck, quack like a duck, or look like a duck - it simply is NOT a duck (this is another profound theory, although not very scientific)!

Now, let's move ahead with the Scientific Study of the Statement that:

GOD IS: A MATTER OF FACT.

Chapter 2 - A Universal Phenomenon

FUNDAMENTAL INTERACTION

Imagine if you will, a Universal Phenomenon that generates forces of unimaginable power; forces that shape heaven and earth; move mountains; part seas. This ever present invisible Phenomenon has, for the ages, continuously influenced human behavior, yet has continued to baffle human description. We have no idea what the Phenomenon actually is; and our best knowledge of how it interacts with the Universe changes with time. This Phenomenon keeps the Universe in equilibrium, allows the planets to maintain their orbits, controls the tides, and influences the weather. This Phenomenon can swallow practically infinite light, or allow it to escape from unfathomable darkness.

This Phenomenon is consistent and predictable; yet its observable effects vary widely. This consistent Phenomenon impacts life and matter in many different ways. For some the Phenomenon can be the firm foundation that keeps them securely connected to their world. For others, the Phenomenon is weak or nonexistent, and seems to have no impact whatsoever on their behavior. This Phenomenon has exactly the same control over individuals or groups; but other opposing influences cause some individuals or groups to react differently under the influence of the Phenomenon.

Now, let's imagine that while man has recognized this Phenomenon since the beginning of our existence, we continue to struggle with a basic understanding of it, or how the Phenomenon interacts with the other things that drive our physical world and human behavior. What are the Natural Laws that are governed by, or respond to this Phenomenon; and why do some of our observations conflict with our current understanding of how this Phenomenon effects us? Finally, does our latest understanding regarding this Phenomenon appropriately describe all the varied reactions that we observe in the Universe?

As much as we try to defy the Phenomenon, or control the effects of the Phenomenon, there is no way we could ever deny that this Phenomenon exists. Have you already guessed what Phenomenon I am referring to?

The Phenomenon is Gravity. How many of you thought I was talking about God?

A Fundamental Corollary. Why start a scientific or academic evaluation of the Statement GOD IS with a discussion on the Phenomenon of Gravity? While they both begin with G, God is certainly not gravity, and gravity is not God.

Gravity is, however, defined as a Fundamental Phenomenon. This is an important distinction, and is the reason that the scientific study of gravity is an appropriate corollary for the scientific study of God. A Fundamental Phenomenon is a foundational principle upon which we build practices, knowledge, and understanding. We use such Fundamental Phenomenon in all branches of engineering.

The phenomenon of gravity demonstrates that there are powerful, undeniable, interactions and effects in our Universe that, while not fully understood, simply cannot be denied.

The introductory statements above purposefully describe a phenomenon without "naming" the phenomenon. In fact, if I had initially called it by its name, "gravity", then the reader may have immediately started to draw certain conclusions without allowing for the discussion. However, by not using a specific name, there was no linguistic limitation on thought. Therefore, we could use as much objectivity as possible in the discussion of the phenomenon, without limiting ourselves to what we traditionally know about gravity.

Of course, now that you know I am talking about gravity, you could (and should) go back and parse the individual introductory statements critically to see if taken as a whole, they describe the phenomenon we know as gravity. As a practical matter, for most of us our learning about gravity is colored by our experience and our point of reference; which happens to be as Earth Beings who have generally completed the normal course of study in middle school or high school science.

The opening sentence of this chapter (for all of you who are parsing) imagines a "Universal Phenomenon", and all statements flow from that. Therefore, a scientific study of the Universal Phenomenon of gravity should start with a discussion of the concept of a "Phenomenon", and view it from variable reference points in its sphere of influence, the "Universe".

WHAT IS A "PHENOMENON"?

We often speak of the Force of Gravity; but gravity cannot be described as simply a force. Gravity is more appropriately defined as one of four Fundamental Interactions of the Universe; the other three being Strong Interactions, Weak Interactions, and Electromagnetism. The accelerating effect of Gravity may create a Force. For example, the force of weight we experience on earth is the effect of gravity acting on a mass in proximity to our planet. However, if we take that same mass to a ship in deep space, that same gravity will have a different effect on the same mass, resulting in a different force. Let's look at this from the simple equation of Force:

$$F = m \times a$$

In this equation, "F" represents force and "m" represents mass. The accelerating effect of gravity is inserted for "a". On earth, this effect is approximately 9.8 meters per second squared. As the accelerating effect of gravity changes, the numerical value for "a" changes, resulting in a different force, "F". The mass stays the same, but the force changes. Of equal importance is that while Gravity is Universal, it's effect; the number we insert for "a"; varies from nearly zero to almost infinity.

Gravity is a Phenomenon. From a scientific perspective, a phenomenon is an observable occurrence, action, or thing. Sometimes, we take phenomenon a bit further, and apply it to an extraordinary or remarkable event, thing, or being - as in a Phenom. Based on what we know about gravity's influence on everything in our Universe, you could say that gravity is a Phenomenal Phenom.

Despite how phenomenal this phenom is we still don't fully understand the phenomenon of gravity. In fact, the concept of gravity has been evolving over many centuries, with definitions dating back to Aristotle and beyond. Early theories were profoundly impacted by further study and revelation by such notables as Sir Isaac Newton and Sir Albert Einstein. Each new theory of this powerful interaction was revolutionary. Since these new ideas broke or completely dispensed with the traditional knowledge of the time, they were initially subjected to criticism.

Einstein's theory of gravitation varied from Newton's theory of gravitation, which in turn differed from Aristotle's theory of gravitation. Three of the brightest scientists/physicists/philosophers of their time; one could argue of all time; developed different theories on the cause of gravitational effect based on their observations, studies, reflections, and Persuasion. More on this term "Persuasion" in a later chapter. For the purpose of this illustration, let's look at the latter two of these theories.

Newton's Gravity. Newton's theory resulted in the Law of Universal Gravitation, which is what most of us learned as the practical equation that gravity is the attraction of two physical bodies directly proportional to their mass, and inversely proportional to the distance that separates them. Hence, space vehicles can escape earth's gravitational pull if they can provide enough separation; and be captured by another planet's gravitational effect, or "pull" as the distance between the two bodies closes. Once disclosed, Newton's concept seemed pretty simple. By the way, Newton published this theory in 1687 in the *"Mathematical Principals of Natural Philosophy"*: Yes, a work on Natural Philosophy.

Einstein's Gravity. Fast forward a little over two hundred years. After Einstein published his Special Theory of Relativity in 1905, he advanced to the next step of determining how the relativistic framework aligned with Newton's now Traditional theory on gravity. Einstein somehow had to reconcile the practical success of Newton's Law of Universal Gravitation, with his own Special Theory of Relativity. He wanted to accurately account for or predict other natural occurrences in the Universe that he had theorized as a result of his observations. Such existences included gravitational time dilation, gravitational lensing, and the fundamental of the space-time continuum; or Spacetime.

As a result he published the Geometric Theory of Gravitation in 1916. This theory established that gravitational effect is a result of matter bending the Spacetime curvature. Einstein also discovered the ten simultaneous, non-linear differential equations, or "field equations of general relativity" that relate the presence of matter and curvature of Spacetime.

I hope you memorized that! Forget about the apple falling from the tree - now we're into serious knowledge. So, based on Einstein's theories and ten simultaneous, non-linear differential equations, we must now have "solid" knowledge on gravity. We can now all agree that we know what gravity is, and how it behaves.

Not so Fast! Einstein's theories did not, it turns out, solve the riddle of gravity. He did move knowledge forward in time, creating a new plateau from which to learn. However, within a very short time frame, as physicists started "reasoning" with the new information, they discovered that Einstein's Special Theory of Relativity pushed logic and math to the limits, literally - Infinity. Einstein also introduced the Cosmological Constant into his equation to deal with a static Universe. When the Universe was found to be expanding, Einstein called the introduction of this Constant a big blunder, and the constant was abandoned. However, Science now theorizes an accelerating expansion of the Universe, so the Cosmological Constant was reintroduced for a new purpose.

Another challenge resulted from the world of quantum mechanics, which observed ambiguity or inconsistency with regard to Einstein's Theory. In order to comport with quantum theory, some Scientists have hypothesized a particle called the Graviton as a Force Carrier for Gravity. Yet, this hypothesis is at odds with what we know about Gravity. So, it appears there is still a lot we don't know about something every one of us know exists.

Despite all of our knowledge of gravity, we still don't know how gravity came to be, what causes it, or what drives it forward in time. Also, all of our theories assume that gravity is universally constant. What happens, however, if gravity decays over space or time; or behaves differently at different locations in the Universe? More importantly, our celestial observations are inconsistent with our current knowledge of gravity unless we add, or "plug in" some unknown or unobserved thing. Enter Dark Matter.

Dark Matter. The hypothesis of Dark Matter was advanced by the Dutch astronomer Jan Oort in 1932 to account for "missing mass" in the Universe. The problem astronomers and physicists faced was that current equations describing the effects of gravity did not comport with what was actually being observed. Convinced that both the equations and the observations were correct, the concept of Dark Matter was introduced to reconcile the differences. After all, either Dark Matter exists, or our observations or present knowledge of how gravity affects the Universe is incomplete; or worse, wrong. Dark Matter has never been observed and its presence is only inferred; yet it is now widely accepted by the scientific community.

It is important to reiterate that Dark Matter has never been seen or detected. It apparently emits no light, electromagnetic radiation, or other discernible energy. We have never observed any direct evidence that it exists. In fact, the only evidence that Dark Matter exists lies in the fact that if it doesn't exist, either our equations or our observations are wrong. Postulating Dark Matter's existence is more palatable than accepting that our current knowledge may be in error.

Dark Matter is not just "rounding error". It is not merely a placeholder for what might be a missing planet or galaxy. Instead, Dark Matter is estimated to account for more than 80% of the Matter in the Universe; yes 80%! Ordinary matter (the matter we can observe or detect) would therefore account for less than 20% of all matter.

Dark Matter isn't the only "plug" necessary to close the gap in our knowledge. Hypothetical Dark Energy fills an even larger equation-observation gap. If you want to learn more about these intriguing theories, you can Google© both of these terms as a starting point for further study.

When it comes to gravity, you can clearly see that while we know a lot, there is still a lot we don't know. In addition, in order to make what we do "know" work, we must infer the existence of things that we don't know, such as Dark Matter. That's why we call Gravity a Phenomenon.

A "UNIVERSAL" REFERENCE

Now that we have examined the Phenomenon of Gravity, let's look at how we observe and understand the Phenomenon throughout its sphere of influence - our Universe.

Our understanding of anything is based on our point of reference, or the situation from which we observe that thing; both physically and mentally. To put it a little more simply, any two of us can look at exactly the same thing and justifiably interpret it differently if we view it from a different reference point. Since gravity is a Universal phenomenon, Balanced Knowledge on the subject should be acquired by interpreting the effect of gravity from other possible reference points in our Universe. If we don't, our knowledge will be based on a single point of reference, and therefore limited.

Most, if not all of the readers of this work experience gravity as land-dwelling beings on a single terrestrial ball, Earth. (If there are any non-Earth dwelling beings reading this, please give me a call. I would like to discuss this further from your perspective). Therefore, our traditional thinking on gravity falls within a fairly tight "bandwidth" relative to the entire Universe.

In our early introduction to physical science, we learned that at or near the surface of the earth, gravity effects all mass equally. As the legend of Galileo's ball drop from the Tower of Pisa taught, if you drop a bowling ball and a tennis ball from the top of a tall building they would hit the ground at the same time. This is a difficult concept to "reason" based on our individual experiences. Shouldn't the heavier ball fall faster? Of course, we know that when we take away the impact of other effects (a hurricane wind for example), and drop our two balls in a vacuum, they will indeed prove what Galileo reasoned. Both balls will accelerate at the same rate due to the effect of gravity.

If, however, you had lived your entire life in the sea, you would have experienced the effect of gravity differently than your corresponding land dweller. If you want to test this theory, go to a swimming pool and drop the bowling ball and tennis ball in the deep end of the pool. Which ball hits bottom first? The effect of buoyancy impacts how you observe the effect of gravity. Now, if your entire experience with gravity had been beneath the surface of the swimming pool, then your observations and understanding would be colored by this reference point. Even if you called your BLF (Best Land Friend) on the phone, you would have a tough time understanding his or her experience with gravity.

Scuba divers adjust their buoyancy compensators to neutralize the competing effects of gravity and buoyancy, creating the effect of "soupy space flight". Many marine animals have developed the natural ability to maintain neutral buoyancy; eliminating the feel of "weight" as a force resulting from gravity. The underwater "atmosphere" is more viscous than that in the above surface world, and therefore observations from this reference point might not support the theory that more dense objects fall under the same accelerating effect as less dense objects.

Likewise, if you had been born and raised on a ship in deep space, never venturing close to any planets, stars, or other massive celestial bodies, you might not even be aware of gravity, having never observed its effect or felt the resulting force. At the very least, you might define gravitational effect very differently. Possibly your spacecraft maintained a "special" gravitational field through its acceleration profile. Or, you may adamantly deny that gravity even exists, having never personally experienced it.

Now that we have examined a couple of macro reference points, let's turn our attention to a micro reference. If you were a happy little proton, you and your quarks would exist in a world dominated by Strong or Electromagnetic interactions; forces that are to the 36th or 38th power stronger than the effect of gravity. Therefore, while you may have heard of gravity from a neighboring neutron, you wouldn't care about or even consider gravity as a factor in your life. The other forces would be so dominant that none of your atomic particle family would even concern themselves with gravitational effect - even though without the gravity dominated outside world, neither you nor they, nor your entire atomic world, would exist.

So you can see that in order to gain a better understanding of Universal gravity, we must examine this Phenomenon from a number of points in gravity's "Universal" Reference.

GRAVITY IS

Gravity has been around since the beginning of the Universe. In fact, the Universe could not exist without the forces created by this Fundamental Interaction. No one denies this fact. We don't talk about Aristotle's gravity, Galileo's gravity, Newton's gravity, or Einstein's gravity. We don't talk about a different gravity on earth and a different gravity on the moon. Fish experiencing "weightlessness" are subjected to the same gravity as the incredibly dense protons residing in the nucleus of atoms.

Our understanding of gravity has continually advanced through the observations and "revelations" of philosophers and scientists over the course of centuries. We have had and continue to have different schools of thought with regard to gravity; different "belief systems" if you will about how gravity behaves. Solid theories of this Phenomenon, taken as fact at the time of their revelation or publication, have subsequently been supplanted by new theories or disproved all together. Our Traditions on the subject are questioned by these revelations and discoveries; which in turn, create "new" Traditional thinking.

Despite all these changes, different observations, and evolving theories, there is only one Phenomenon we call Gravity. When our knowledge on the behavior of gravity is contrary to our observations, or even disproved, we don't question whether gravity is real, or deny its impact on the Universe. Science will even go so far as to infer the existence of significant unseen and unknown things just to support our current understanding of gravity.

Whether you live in the sea, were raised on a spaceship, thrive in a subatomic galaxy, or just live like the rest of us on good old Planet Earth, gravity in fact exists. It was, is, and always will be a Fundamental Interaction in the Universe. Even if you don't recognize it, choose to ignore it, decide to deny it, try to defy it, or allow other forces to mask its presence; Gravity IS.

GRAVITY IS a MATTER OF FACT.

GOD IS a MATTER OF FACT.

Chapter 3 - Why Knowing Matters

WHY THE FACT MATTERS

GOD IS a matter of Fact - but why does the Fact Matter? Why would anyone write an engineering treatise on the statement GOD IS? Coincidentally, why would anyone read a hundred and fifty pages or so on such a simple yet profound statement? Further, if most people know that GOD IS, why do we need to explore the question from a scientific perspective?

Again, our answer lies in two parts: **Restricted Science** and **Human Interaction Observations**.

RESTRICTED SCIENCE

We introduced this bias in the first chapter, but additional discussion on the very important impact of Restricted Science is warranted.

Less than two hundred years ago some of the brightest minds of the scientific community declared that God was dead - or at the very least, irrelevant. Scientific discovery in the previous three centuries had expanded man's knowledge of the natural world rapidly, and Scientists quickly displaced both Philosophers and Theologians as the leading thought merchants for issues ranging from the beginning of the Universe to the meaning of Life. Scientific Experimentation and mathematical logic quickly overtook both Tradition and Reason combined in providing the theoretical and practical solutions to all human question and concern.

From the late 1800's into the early 1900's, as confidence in scientific knowledge grew stronger and bolder, an emerging preferential School of Thought was that Scientific Method was the all powerful tool; and that logic, experimentation, and observation supplanted all other foundations for knowledge. Leading thinkers across the scientific board, from Physicist Paul Dirac to Neurologist and Psychotherapist Sigmund Freud, had little or no use for previous Scientific, Philosophical, or Traditional concepts of God; setting aside any notion of or need for a higher level of Intelligence in a world dominated by man's Scientific Knowledge.

By the year 1900, scientific experimentation and observation were becoming the only acceptable standard for the development of all "Reliable Knowledge". New pronouncement of fantastic scientific discovery, even highly speculative theories, became fashionable; as physicists, psychologists, astronomers, geneticists, chemists, and biologists became the new rock stars of philosophy. In a few centuries the roles of the two subjects had reversed. In Newton's time, scientific thought emerged from philosophy and reason, as in his famous work, *Mathematical Principles of Natural Philosophy*. Now, however, philosophy and reason are often discarded by a selective group of highly influential theorists who argue that the reliability of their scientific experimentation and the precision of their mathematical logic trump all other sources of knowledge.

Under such Restricted Scientific thinking, proponents of "godless" concepts for the initiation of the Universe and the Creation of Life have proclaimed superiority to other avenues of learning. Nothing is more indicative of the perceived manifest superiority of Restricted Science than Stephen Hawking's proclamation that "Philosophy is Dead" at Google's 2011 Zeitgeist Conference. Hawking stated that:

> "Why are we here? Where do we come from? Traditionally, these are questions for philosophy, but philosophy is dead."
>
> - As reported in *The Telegraph*, May 17, 2011, by Matt Warman, Consumer Technology Editor

Dr. Hawking went on to say:

> "Scientists have become the bearers of the torch of discovery in our quest for knowledge."

In just over a century, the school of Restricted Scientific thought, represented by a select group of charismatic scientist-philosophers, has declared both God and Philosophy dead.

Ironically, science and math are both fallible. Scientific Method and Mathematical Logic are just two of the tools of Learning that allow us to expand our Balanced Knowledge of the World. However, neither tool is perfect, and both are subject to error. Yet, despite the fact that these modern day scientists-philosophers "disproved" previously reliable and useful scientific laws, they believe their own knowledge to be indisputable; their new science "correct"; their theories "infallible"; and therefore, God and philosophy no longer necessary.

Unrestricted Science is a good thing. This is not an argument against science. Quite the contrary, Scientific Method is a wonderful tool in the learning arsenal. Scientific achievements in a wide range of subjects; from medicine to economics, from chemistry to genetics; have significantly improved the Human Condition and our understanding of our Universe. Science and Scientific Method however cannot be controlled by or restricted to a single school of thought. Scientific analysis must be open to broad discussion and critical thinking. Scientific Method is not solely the province of Theoretical Physics, Cosmology, or Quantum Mechanics. Further, scientific experimentation and mathematical logic must be analyzed in the light of other learning through Tradition and Reason.

For some reason, we continue to allow a portion of the scientific community to insist that only such "godless" scientific theories are valid; without allowing equal time for a discussion of the clear evidence demonstrating a contrary theory; that God exists. We allow science to teach that God Isn't but won't allow science to explore the alternative that God Is! This Restricted Scientific thinking is in itself, contrary to Scientific Method, which should allow for rigorous analysis of both points of view.

HUMAN INTERACTION OBSERVATIONS

Our need to allow for the scientific study of God becomes increasingly apparent in light of the observations of Human Interactions across four different Grand Social Science Experiments: Marxism-Communism, Nietzsche's Ubermensch, Restricted Theocracy, and Tolerant Democracy.

Marxism: Karl Marx is considered one of the brightest and most accomplished thinkers of the 19th Century. Marx is credited with advancing the concept of Social Science. Indeed, Marx insisted that Scientific Method and logical analysis could in fact accurately predict Human Interaction on a large scale.

Two of the underlying influences in Marx's development were his atheistic view toward God and his involvement with the radical thinking Young Hegelians. As part of that group, Marx attacked religion in an attempt to undermine the Prussian establishment, ultimately referring to religion as an "opiate for the masses." Marx asserted that "man is the highest being for man". His ire cycled between religion and capitalism as the corrupting power that the Prussian aristocracy used to control the working masses.

Marx's theories were exquisitely framed, and his work quickly influenced a wide audience. Marx considered himself a "scientist", insisting that his model of perfect social balance was achievable if the corrupting influence of Capitalism and the controlling influence of Religion could be removed. Internal tensions would drive a socioeconomic system to Socialism, which in turn would evolve to a truly "classless society" called Communism.

This brief discussion barely scratches the surface of Marx's Theory, but suffices to demonstrate the evolution of thought that was the basis for a form of government that, during the 20th Century, was used to attempt to control nearly a third of the world's population. Of course, the results of Marx's scientific theory of "godless" Human Interaction in a classless, conflict free society, has been calamitous. The Soviet Union disintegrated, while Soviet puppet states such as North Korea and Cuba are remaining examples of exactly what such "godless" Social Science can produce. Even China has had to rethink the strictly applied Marxist theory, and has significantly relaxed the strangle hold on individual thought in order to allow some societal equilibrium as the Chinese people explore new boundaries in personal and social freedom.

In less than one hundred years, Marxist theory of "godless" Social Science, including his complete disdain for Capitalism and Religion, has been significantly discredited; yet Marx is still considered one of the principal architects of modern social science.

Nietzsche's Ubermensch. Another famous Social Scientist-Philosopher of the 19th Century, Friedrich Nietzsche, had as one of his central themes, "God is Dead". While there is some debate about what Nietzsche may have meant, this clearly articulated philosophy in his work *The Gay Science*, created an underlying foundation for much of his theory. Coincidentally, Nietzsche, like Marx was born in 19th Century Prussia, and was also highly influenced by the social imbalance between the Prussian aristocracy and what he saw as the plight of the common man. Hence, just as with Marx, Nietzsche used the strength of scientific study to validate his case for new thinking in Social Science.

Much of Nietzsche's thinking rises out of his tendency toward Nihilism; a philosophy that generally argues that life is without intrinsic value or purpose – in other words the meaning of Life is Nil. With a Nihilistic mindset, and rising from his declaration of the death of God, Nietzsche offers another of his important concepts, the "Ubermensch"; translated as the Overman, or Super Man. With no God, the Ubermensch became the new creator of core values; the leader of Humanity.

Nietzsche declared that, "The Ubermensch is the meaning of the earth."

Nietzsche saw the Ubermensch as representing the super race of humans; the natural result of the survival of the fittest. Just as with pure Capitalism, Nietzsche theorized that the natural competition to survive and even thrive would result in the emergence of a superior race or Ubermensch, who could and should control the course of Human Endeavors. With God dead, the Ubermensch was the new god. Nietzsche believed that the Super Race provided the best course for human interaction and advancement - even if such advancement was predicated on chaos, war, and destruction.

Rightly or wrongly, Nietzsche's powerful argument struck a cord in early 20th Century Germany, serving as the foundation for much of the inflammatory rhetoric of the Third Reich, and the Fascism of Nazi Germany. Hitler used Nietzsche's concept of the Ubermensch to form his own rationale for the master race that was destined to control the world.

As with Marxism, social and political systems based on Nietzsche's godless Ubermensch concept proved disastrous. The replacement of God with the Ubermensch of Nazi Germany resulted in fanatical nationalism and the most horrific war our world has ever known. I think you could classify this approach as a failure! Yet the concept continues to appeal to those who feel their social or political strength threatened, and would use racial purity concepts as the principal for social order. Nietzsche's philosophical concepts and principles are still widely studied.

Restricted Theocracy. Throughout the world, in cultures that continue to be dominated by either a single or group of tightly aligned Religious Traditions, the scientific study of God as a Universal phenomenon is often highly discouraged, if not completely disallowed. Such scientific study may call into question or criticize the singular dogma dictated by a Religious Tradition, encouraging alternate theories, and creating instability in the absolute control of the Theocracy. Restricted Theocratic regimes control their subservient populations by claiming to possess special knowledge, connection, power, or authority from God. Scientific evidence that undermines religious dogma, practices, or historically defining moments may call into question the validity of the authority upon which the Theocratic leadership stands.

Europe saw the results of just such a collision of Religion, Philosophy, and Science when Copernicus determined that the earth orbited the sun; shaking the Tradition concept that the Earth was fixed at the center of the Universe. Centuries of struggle ensued as the Age of Enlightenment wreaked havoc on a broad range of

Tradition concepts. Yet, the overall Human Condition improved as the Balanced Knowledge acquired through Reason, Experimentation, and Tradition evolved.

If we fail to study God's existence and examine God's influence scientifically, we allow Restricted Religious Tradition to be the dominant or absolute foundation for Human understanding of the powerful forces that flow from the existence of God; specifically the fundamental influence of Supreme Intelligence.

Tradition is a very important part of our Learning Experience, and supports an appropriate Balance of Knowledge. Tradition is a good thing, particularly when combined with Reason and Experimentation. However, Restricted Tradition as the single source of knowledge, or as the single basis for our understanding on any subject, can prove to be very Intolerant. It also has a spiraling effect, for as we attempt to defend Tradition against the onslaught of Reason or Experimentation, we become more entrenched in our own singular understanding. Hence, we become more defensive, more intolerant. This type of learning often results in parochialism and partisanship; or even worse, fanaticism.

We can see this interaction at work in the extremist actions of some religious fundamentalists. Under a highly restrictive Theocratic environment, with only one view of the power of God, a closed society can be quickly turned into a radical mob; intent on protecting the singular theocratic world view from the incursion of other thinking.

It is imperative that human knowledge of God not be restricted to singularly isolated and radical schools of thought. Only through a full and open discourse on God, and the impact of God on the Universe and Human Interaction, can we find common ground upon which to build our social structures and allow for more tolerant Human Behavior.

Tolerant Democracy. In the United States, we talk a lot about the underlying fairness of the principles of Democracy. However, this takes a simplistic view of the strength of the Tolerant Democracy that our Founding Fathers created. A Tolerant Democracy is not a pure Democracy.

In a pure Democracy, the Rule of Law is made by majority vote of the people. That sounds fair and simple. However, such a government makes it very easy for the majority to trample on the rights of the minority or the individual. Therefore, if a democracy is to provide fair governance for all citizens, it must secure individual rights, encourage minority views, and tolerate dissent. The framers of the U.S. Constitution recognized that while democratically elected representatives should establish the Rule of Law, that any such law or regulation should not and could not take precedence over the "Blessings of Liberty" to which each citizen is entitled.

The United States was founded based on the concept that people are entitled to certain Unalienable Rights that flow from a higher intelligence; a higher authority; the Creator. With regard to the concepts for governance of Human Interactions, one of the most famous and empowering phrases ever written comes from the U.S. Declaration of Independence.

> "We hold these truths to be self-evident, that all men are created equal, that they are endowed by their Creator with certain unalienable Rights, that among these are Life, Liberty, and the Pursuit of Happiness."

The authority and sanctity of Individual Rights as a foundation for the rules of Human Interaction is of immense importance. Majority rule without protection for individual Liberty and tolerance of minority opinion is a dangerous form of government. Fortunately, our Founding Fathers recognized this fact in the Constitution as Amended by the Bill of Rights. This essential principle is firmly established by the recognition of the "Blessings of Liberty".

Blessings of Liberty. The U.S. Constitution states that Liberty is a function of Blessings. Neither of these terms was accidental, and both were carefully crafted to define the guiding principle of our Nation. The impact of this phrase is unquestionable. Human Liberty is dependent upon and a function of Blessing. This was a revolutionary Scientific Theory that our Founding Fathers advanced more than 220 years ago. So let's look at this Theory carefully and fully understand the "Blessings of Liberty".

Liberty is a tremendously powerful term. It's not just about getting to do what you want to do, as long as you don't negatively impact what another person has the right to do. The term is much greater than that. Liberty allows us to think, believe, discuss, and act upon our own feelings, reason, or knowledge. Liberty recognizes that Humans have Intelligence, and can act on that Intelligence through Free Will. Liberty allows us to exercise Free Will as we seek our greatest aspirations.

Of greater importance is the term Blessing. This term is equally powerful, but even more profound in its clear demonstration of the Founding Fathers intention and meaning of this statement. Any number of terms could have conveyed Liberty to the citizenry; such as a grant, benefit, entitlement, or gift. They specifically chose the term Blessing. They didn't restrict Liberty to something the majority conferred, the social order provided, or the Ubermensch awarded. Liberty isn't chattel; it cannot be bought or sold.

The framers of the Constitution used the word "Blessing" because of its clear meaning. Blessings come from a higher authority; what the Declaration of Independence termed as Nature's God. Therefore, the Blessings of Liberty are not subject to the whims or grace of men or governments. The Blessings of Liberty are a birthright of Humanity.

At the same time, our Founding Fathers also recognized that God exists without regard to any specific or restricted religious tradition. Every individual is free to follow any religious dogma or spiritual tradition they choose, or they are free to follow no belief at all. This dual principled foundation allowed for the appropriate Authority for unalienable rights while preventing the intolerance of a mandated religious tradition or Restricted Theocracy. It is important to note the clear separation between the acknowledgement of God's existence, and the religious traditions humans use to find a personal connection to, define a relationship with, or better understand God. God's existence is not dependent upon any religious tradition.

The United States Experiment has created a Human collective or sphere of influence, wherein multiple traditions and cultural practices can come together harmoniously and build on their commonality, instead of dividing on their differences. This same system that provides for a complex social interaction continues to allow individuality and personality in our actions, expressions, and beliefs. These principles have allowed our Nation to evolve from the terrible tradition of slavery and bigotry; and even survive a massive Civil War that tore at every part of our fabric.

The United States success has not been the result of an accident or dumb luck. Just as importantly, the American Experiment was not a result of man's "natural" ability to get along with each other. Quite the contrary, as we know from the above discussion, Marx and Nietzsche both relied on man's "natural" behavior as the basis for their Social Science; Behavior which either sacrificed individual freedoms and opportunity for the good of the social order (Marxism), or subjected freedom and opportunity to the direction, influence, or will of a dominant culture or race, or the extreme alpha personality, the Ubermensch (Fascism). We see expressions of both godless social systems throughout the animal kingdom, from a colony of bees to a pride of lions.

Why has the American Experience succeeded thus far? How can we ensure that our Foundation remains strong? The answer lies in the strength of the two very clear foundational principles introduced as the basis for America's governance of Human Interaction. First, our Founding Fathers recognized that certain Unalienable Rights were endowed by a Creator. Second, and of equal importance, they understood that God's existence and authority were not dependent upon and should not be controlled by any prescribed religious tradition. While America clearly recognized the authority of God in establishing the Blessings of Liberty, people would be free to follow any belief system they wanted, including no belief system at all. As the Declaration of Independence recognized, Human Interaction should be governed the "Laws of Nature and Nature's God".

STUDYING THE FACT MATTERS

We will see through this Engineering Treatise the immense Power that flows from the fundamental nature of Supreme Intelligence.

Man has attempted to control Human Interactions by controlling the Forces of Intelligence through godless social science concepts as well as highly Restricted Theocracies. These experiments in social governance and Human Interaction have been unequivocal failures.

The United States, on the other hand, was built on a strong foundation acknowledging the Blessings of Liberty and unalienable rights endowed by the Creator. For the Rule of Law to be fairly applied, we must recognize the Forces and Behaviors that are fundamental to Intelligence. How is it that in public schools students can be taught the importance of "certain Unalienable Rights", but are not allowed to study the Phenomenon cited as the source of those rights, the "Creator"? What does it say about us when we champion Liberty, but disavow the Blessing from which Liberty flows?

Today, the Restricted Scientific Study of a godless initiated Creation of the Universe and Life is widely accepted in public schools; while the equally valid and compelling Scientific Theory of the fundamental nature of a Supreme Intelligence cannot even be offered, much less studied. A Balanced Knowledge of God developed through scientific study is essential to a full understanding of the Origin of the Universe, the Creation of Life, and the complex Forces of fundamental Intelligence that drive Human Behavior and Human Social Interaction.

GOD IS: A MATTER OF FACT.
SCIENTIFIC STUDY OF THE FACT MATTERS!

PART TWO: How Do We Know?

Chapter 4 - Learning to Know

KNOWING THROUGH LEARNING

Knowing. How can we know that GOD IS? For that matter, how can we know that Gravity IS? That may sound like a stupid question, but how do you KNOW that Gravity IS; or what Gravity Is?

For most of us, the simple answer is that we were told what Gravity Is. However, as we discussed previously the body of human knowledge about gravity is constantly changing.

Honestly, how many of us if asked, would say that gravity is a result of the curvature of SpaceTime? How many of us even know what that means? Do we fully understand the hypothesis of Dark Matter required to allow all of this to make sense? So how do we Know what Gravity Is or that Gravity Is? For that matter, how do we Know that Anything Is? This was in fact a philosophical challenge a century ago, when Einstein's theories upset Newton's applecart (pun intended).

Think about it. Einstein's Special Theory of Relativity, and subsequent Geometric Theory of Gravitation, called into question rock solid science. Centuries of engineering and discovery had relied on Newton's Theory of Gravitation. Therefore, if Newton's science was wrong, despite hundreds of years of teaching and reliance, how could we ever know what's right?

How can we rely on anything? If all knowledge is fallible; and therefore subject to refinement, modification, or a 180-degree change; how do we know anything?

The answer is that we must Learn, and then continue to Learn.

That may sound simplistic, but for some reason the human need to continuously Learn in order to Know is often forgotten. In fact, our current education process particularly in the areas of math and science seems contrary to this fundamental principle. We have become all too finite and mechanical in our Knowing - as if Learning is an assembly line process.

We like to teach by having students memorize things in a timed sequence and an orderly process. This week we memorize "X". Then we test students' memory with a test of "X". If they pass the test, then they are deemed to "Know X". Next week, we move to "Y". Memorize Something - Test Something - Know Something: done!

This process of teaching has nothing to do with the way humans "Learn" or "Know". If we memorize a bit of knowledge, and then test our memory of that knowledge, the only thing we've tested is our memory. We have not tested knowledge. What we "Learned" yesterday, if not applied, refined, and possibly even relearned; can be either forgotten or made obsolete tomorrow. Learning is a continuous activity - a never-ending journey. Therefore, one of the essential elements in Learning is learning how to Continually Learn.

Learning. From great minds such as John Locke, we know that we are born knowing nothing. At birth, our minds' knowledge is a blank paper. Certainly we have instincts, as most animals do. However, from day one, humans begin replacing the instincts required to survive (such as crying or sucking), with the knowledge required to survive; if not to thrive. Further, the only way we acquire and maintain knowledge is through Learning; and the only way to Learn is through Experience. That bears repeating - the only way to Learn is through Experience!

That last sentence may have provoked a challenge or loud outcry, if not a downright denial. Engineers like to look at things in digestible chunks. So for my defense of this statement, "We Only Learn through Experience", I will start with a simple, fundamental Axiom of Experience.

Axiom 4.1: Experience is what we are exposed to over time.

Profound? Not really. Putting it another way, whatever we are exposed to over time is our Experience. We may even restate this in more "scientific" vernacular; Experience is a distance we travel in the time dimension, as measured in minutes, hours, months, or years. We therefore refer to a day's Experience, or a lifetime of Experience.

The next step in defense of this statement is to offer the hypothesis that the only way we Learn is by moving through Time.

This is different from the way a machine "learns". A machine such as a computer can learn a great deal of information instantaneously. We can load a computer program onto a hard drive, and immediately the computer has "learned" both the data that has been stored, its Memory; as well as how to use that data based on the algorithms or logic contained in the software the computer has access to, its Application.

People however, are not machines, and cannot instantaneously store or memorize a load of data, nor can we instantaneously learn how to apply what we know. The only way people can Learn is through the passage of time. The time distance of a Learning Experience may be extremely short, and therefore approaching instantaneous. You may glance at something quickly and retain a memory of the knowledge it conveyed. However, the glance and storage of information is in itself a brief Learning experience.

Touch a hot stove, and you will almost immediately Learn something, assuming you can either feel the heat or smell burning flesh. However, the experience of touching the stove, receiving a pain stimulus, reacting to the user input (jerking your hand back) and storing the data for future use (don't touch the hot stove again) takes some amount of time - it takes the Experience. Furthermore, if you touch a stove that has been turned off for a couple of hours, your actions will result in a different experience. The same action would have taught two different things, because the experiences were at different times.

Reading is a Learning Experience. What you Learn from any reading experience may have as much to do with the environment in which you read, or your frame of mind while reading; as it does with the content, style, or presentation of the material you are reading.

This hypothesis is further evident by the undeniable fact that if humans do not advance in time, then they cannot Learn. Hence the hypothesis is confirmed, and we can offer the following Axiom of Learning.

Axiom 4.2: Humans only Learn through Experience.

THE LEARNING EXPERIENCE

We Know through Learning, and we Learn through Experience. Therefore, to answer the question "How do we Know?" we shall examine Experiences through which we Learn.

Again, taking a relatively simple approach, we will categorize Learning Experiences under three specific areas, or Channels; Tradition, Reason, and Experimentation. Think of these Learning Experience Channels as three different television news channels. One news channel may focus on a historical perspective and commentary; another may provide talking heads that provide a variety of reasoning on a subject; while a third channel may focus on just showing footage of actual events, allowing the viewer to observe and determine what they believe they are witnessing. A single event covered by these three news channels would lead the viewer to three different Learning Experiences.

Many readers will immediately want to expand on my short list, but for the sake of argument (and simplicity), we shall start with these three Experience Channels and see if any Learning gets omitted. Further, I am specifically eliminating "instinct" as anything we learn, since instincts are defined as behaviors that are performed in the absence of learning.

Tradition. The Tradition Channel includes practices, principals, knowledge, dogma, customs, actions, events, and other behaviors previously experienced and believed to be true, factual, accurate, relevant, or valuable; and therefore worthy of being retained in some form and transmitted through time so that others can realize the benefit of the Experience.

We often think of Tradition as practices from past generations, but that's not always the case. For example, many of us have our own Holiday Traditions, and we like to have those experiences over and over because they generally lead to a more predictable outcome. If our children enjoy, or "value" these Holiday Traditions, they will carry it through their Experience, passing it along to their children.

Tradition is the body of learning acquired by previous Experiences, whether those Experiences are our own or those of others before us. Traditions are typically transmitted or passed along through generations or cultures, until they are no longer valued, for whatever reason, by those learning them. At that point, the Tradition is broken or discarded, typically being replace by new Tradition.

Tradition is a starting point for most of our Learning. Tradition eliminates the need to reinvent the wheel. If we didn't have Tradition, each of us would have to start at the beginning, learning everything from our own personal Reason or Experimentation, without the benefit of all of the Experiences of those before us.

A significant portion of our early formal education is based on Tradition; not only what we learn, but also how we learn. Our learning is based on Traditional subjects experienced through Traditional teaching methods. Our textbooks are based on Tradition, as are many of the stories, lectures, examinations, and teachings. Clearly learning in history and philosophy are based on Tradition. Your knowledge of the War Between the States, the American Civil War, or the War of Northern Aggression (all of which represent the same historical conflict) may be colored by the Tradition influences of your geographical location or culture.

Are you ready for another eyebrow raising statement? Tradition also influences learning in both Science and Math; the two areas we generally look to for more "precise" or reliable knowledge. Science and Scientific Method are clearly driven by cultural Traditions that tend to influence how we experiment or observe, or how we accept the experimentation or observations of others. We often accept or reject science based on our Traditional valuation of the science being considered.

A perfect example of this is acupuncture. Acupuncture is an ancient form of Traditional Chinese Medicine, having been practiced for centuries as effective treatment for a variety of ailments. However, while this is Traditional Science for Chinese cultures, it is "Alternative Medicine" in western cultures where it is non-traditional, and has had difficulty being widely accepted by the western scientific community.

Tradition may even color the precision or valuation of math concepts. Our understanding of math expressions can be subject to the limits of our Tradition.

Let's start with a simple concept: **1+1=2**. That expression is "exact" under our Base 10 Tradition. However, if our Tradition was binary, or Base 2, then that expression would be wrong; instead we would state that **1+1=10**.

We may take this a step further with the factual statement that **6+6=C**. That expression is equally as accurate as **1+1=2**, and **1+1=10**. If you grew up immersed in "computer" Tradition, all of these mathematical expressions are accurate and make sense. You would immediately recognize **6+6=C** as a hexadecimal notation; wherein **F+F=1E**. However, if you are exposed only to a Base 10 Math Tradition (as most of us are), then you might find two of these mathematical expressions confounding; or worse, you may actually deny the accuracy of these statements altogether and even argue that they are wrong.

Tradition anchors our knowledge. However, as with any anchor, it can also keep us from moving ahead if we constrain our Learning to this single Channel.

Reason. The next Channel of the Learning Experience is Reason. Reason is our own mind's Learning Experience. We may read, study, and even subscribe to the reasoning or philosophy of others, but that isn't our Reasoning. Accepting another person's reasoning, or a body of reasoning, without conducting our own Reasoning, is a Tradition Learning Experience. The only way we learn from our Reasoning is if we ourselves conduct the Reasoning. I realize that this sounds obvious. However, a major societal challenge occurs when a significant number of the members of a society cease doing their own Reasoning, relying primarily or even solely on the Reasoning of a select few.

We see this phenomenon every day in the illogical "knowledge" that gets passed around through the Internet. I get numerous emails from well-intentioned contacts who pass along new knowledge or information that simply does not pass the old "sniff" test - the test of Reasonability. Therefore, we may need several additional data points to either verify such new information, or to confirm our Reasonable suspicion of its accuracy. We are each blessed with Intelligence, and are called upon to use our own Reasoning in order to filter the flow of information passed to us.

This is not to say that the reasoning of the past, handed down as Tradition, is not valuable. However, all of the reasoning passed down through Tradition and incorporated in our Learning does not substitute for our own Reasoning. We need to exercise our own personal Reasoning as part of our Learning Experience. In fact, absent Reasoning, even a highly educated society can be led astray by the bad Reasoning of a few. The rise of the Third Reich, including the philosophy of Adolf Hitler and the resulting atrocities of Nazi Germany, is a clear example of the failure of a highly educated society to apply their own Reasoning against the arguments of a handful of wicked, charismatic leaders.

Complex, or multi-stage Reasoning is a definitive characteristic of Human Nature. It is one of the fundamental Forces that is initiated by Intelligence. Through Reasoning we can foresee cause and effect, truth and falsehood, good and bad. Reasoning allows us to question, doubt, and critically apply Knowledge.

Reasoning must also be practiced, or exercised. We can call on Tradition to learn how others before us reasoned and how they applied reasoning to address problems. We can then begin to formulate our own Reasoning, including how to use this exceptionally powerful Experience Channel.

The Reasoning Channel forms the basis for Critical Thinking; the application of what we have learned to determine appropriate responses to questions or challenges. In other words, we must Learn to Reason, and then we must continue to Learn from our Reasoning.

Experimentation. The final Channel of Learning Experience is discovered through our own personal Experimentation. For the purposes of this discussion, I am using the term Experimentation to describe the entire process of observation, hypothesis, testing, data collection and analysis, and finally conclusion.

Experimentation is a primary tool for the Scientific Method. The term Experimentation is closely aligned with Experience, and is the category through which we Learn by physically doing something or personally observing something; extracting Knowledge from our own activities; our own trial and error.

Experimentation can be simple actions we take on our own to either prove or disprove something. This is commonly observed in the juvenile actions of children, (and the juvenile actions of some adults) who are often bound and determined to challenge the information passed along to them through Tradition.

Let's return to the hot stove discussed previously as a young scientist at the ripe old age of five ponders a "burning" question. His mother has told him not to touch the stovetop or he will get burned. Yet, he sees mom touch the stovetop often, even cleaning it with thin paper towels. To date, his Knowledge of the result of touching the stove is based on the Traditional information given to him by his mother, and the Experimentation he observes when his mother cleans the stovetop. He Reasons that these two Learning experiences are contradictory, since his mother did not burn her hand. He therefore determines that the only way he can Know the truth is to conduct his own Experiment - he must touch the stove!

Now, here's the problem. This simple Experiment can end in one of two ways - either of which will create a conundrum in the young scientist's mind. If he touches the stove and gets burned, it confirms the Tradition passed down to him by his mother. He should have listened! However, he is now left with the question of why his mother does not get burned when she touches the stove?

If on the other hand he touches a cold stove, then he has discovered that his mother's Tradition is wrong, and therefore is of no value to him. He can in fact touch the stove without getting burned! He now begins to question other Traditions mom is attempting to pass along. If mom was wrong about this, what else?

Experimentation can go to the other end of the spectrum; where large teams of scientists, physicist, mathematicians, and engineers will conduct experimental activity over years, for billions of dollars, in an attempt to prove a theory or model.

Have you heard of the Higgs Boson? Physicists have a theory called the Standard Model that deals with the relationship of Strong, Weak, and Electromagnetic Interactions, and it is now widely accepted as Traditional science for particle physics. However, if the Standard Model is correct, then the Higgs Boson exists. For that reason, there is an all out effort to find evidence of the Higgs Boson, thus confirming the Standard Model.

In this case, the experimental activity is a continuous series of experiments that form a body of Experimentation Learning. Note here that the lack of data proving the theory correct does not mean that the theory is incorrect. Instead, the scientists continue to adjust, modify, or refine the individual experiments in search of data that will in fact, prove them correct. By the way, evidence of the Higgs Boson appears to have been confirmed in the summer of 2012; but there is still much testing that will be conducted (and money to be spent) to shore up this confirmation.

Sitting somewhere between the above two extremes, most Experimentation Learning evolves from a single person's Experience, through which she or he continuously conducts a series of experiments in an attempt to confirm a hypothesis; until something is proved or disproved - possibly even by "accident". A crystal clear example of this type of Experimentation Learning is in observing how many people "experiment" with texting while driving.

Tradition Learning, passed along to us from the Experience of others, tells us that texting while driving is a high-risk activity that may result in serious injury or death; or at least a much higher probability of serious accident. Yet, many drivers discount the value of this Learning based on their own Reasoning and Experimentation.

From observations, some drivers Reason that they should be able to safely text while operating a motor vehicle. Therefore, they decide to conduct a personally controlled and "safe" experiment to confirm this hypothesis. Stopped at a traffic signal, they whip out the cell phone and send a quick text. The initial experiment is completed, data collected, hypothesis confirmed. The result of this initial exercise in Experimentation Learning contradicts the previous Tradition Learning Experience. The driver not only safely texted while behind the wheel, but also felt safe and in control during the process - they were after all, sitting at a light.

Naturally, the driver is encouraged by this success. Feeling emboldened, they continue the Experiment, and are soon texting in slow moving traffic, glancing up regularly to "text safely". Pretty soon they have moved on to interstate texting, keeping one eye on the windshield and another on the cell phone display.

The driver's Experimentation Experience has displaced the Tradition Experience - they have "Learned" that they can safely text and drive. Reasoning and Experimentation allowed them to move beyond older and outdated Tradition; what others before had learned and attempted to convey. They were able to collect their own data, confirm their own reasoning, and prove to themselves that in fact, they can safely text and drive.

Their Experimentation freed them from the shackles of Tradition and "old fashioned thinking", and …. Screeeech, BOOM. They were just killed in a head on collision texting while driving.

BALANCED KNOWLEDGE

Knowledge is what we Learn, and Learning comes to us over time, through Experience. Further, we Learn through three separate Experience Channels: Tradition, Reason, and Experimentation. Knowledge can therefore be metaphorically defined as a three-legged stool. In order for our knowledge stool to support us, the length of the legs must be approximately equal.

This phenomenon requires the development of Balanced Knowledge, which is the most reliable body of knowledge upon which we can individually or corporately rely. All three Channels are critical, and no single Channel is more or less important than the other two. Taking this one step further these Channels cannot be isolated one from the other, as each Channel is an integral part of the other two Channels: To wit Tradition and Reason shape our Experimentation; Reason and Experimentation shape our Tradition; and lastly, Experimentation and Tradition shape our Reason.

The legs of our knowledge stool do not have to be exactly equal for the stool to balance; in fact, they probably never will be precisely the same length. The relative lengths of the legs are always changing, as we move through time; through our experiences. The challenge is to keep the differences in lengths within a manageable range.

However, if we allow the majority of our Learning to flow from a single Experience Channel, one of the legs of our knowledge stool will inevitably begin to dominate our thinking. Our seating becomes less firm; our foundation less stable. Left unchecked, this instability causes us to cling more tightly to the single Experience Channel, whether Tradition, Reason, or Experimentation, as our sole reliance for Learning. At some point, our knowledge stool will begin to list badly.

Again, let us consider our news channel metaphor. If you get all your information from a single channel, then the perspective or philosophy of that single channel will establish the lens through which all your news is colored. At some point, that channel's perspective will become your perspective, no matter how prejudiced or unbalanced the reporting may be.

The knowledge stool is NOT, unfortunately, self-righting. A single dominant leg can quickly spiral out of control, knocking Balanced Knowledge right out from under us. The results can be catastrophic, for both the individual and society. Single Channel Learning is what leads to intolerance, extremist views, and fanatical behavior.

Therefore, we must constantly strive to keep all of our Learning Channels open. We need to be acutely aware when a single Channel begins to dominate either our individual or societal Balanced Knowledge. Our Learning, public and private, needs to incorporate an appropriate measure of Tradition, Reason, and Experimentation in order to appropriately Balance Knowledge.

The Way we Know anything is through Balanced Knowledge resulting from Continuous Learning through all Experience Channels.

BALANCED KNOWLEDGE ALLOWS US TO KNOW THAT GRAVITY IS.

BALANCED KNOWLEDGE ALLOWS US TO KNOW THAT GOD IS.

Chapter 5 - Scientific Persuasion

SCIENTIFIC METHOD

We marvel at Science. Whether pondering the expansiveness of our Universe or the intricate workings of nano-technology, scientific study seems to open our mind to new possibilities. Clearly, scientific advances in everything from health to nutrition, from consumer products to transportation; have contributed positively to the Human Condition. Scientific discovery touches every aspect of our lives; Science addresses everything we do. Science is a good thing.

But science is not an isolated subject left unto itself. The term comes from the Latin *Scientia*, which simply means "knowledge". Some definitions refer to science as a body of reliable knowledge. Science is however more than information to Know. It is in reality a method of Learning, which of course is the key to acquiring and maintaining Balanced Knowledge. Scientific Method is a description we generally use for the Learning we acquire through the Experimentation Channel; using observation, experimentation, and documentation to establish an indisputable (or at the least, a widely acceptable) foundation for a theory, postulate, or statement that becomes Reliable Knowledge.

Science; or more specifically, Learning through Scientific Method, is however a two edged sword. Some may argue that scientific advancement comes with its own set of challenges. With every new discovery, we are often faced with new "unknowns". For each question we answer, two more questions pop up. That's the Paradox of Knowledge: The more we know, the more we know we don't know.

So why do we choose to know? Unfortunately, this paradox often causes us to want to put on our blinders. We sometimes choose "not to know", as a remedy to knowing that we don't know. Do you feel like you're on a roller coaster? In other words, sometimes it's easier to bury our head in the sand than to look at the challenges we face.

Science and Scientific Method, are good things. Choosing not to know is clearly not in our short term or long term interest. The benefits of scientific discovery outweigh the negative impacts of such newly found knowledge. Human pursuit of greater knowledge is in our DNA. Unlike the rest of the animal kingdom, humans are not satisfied with simply existing on Planet Earth for a short time and expiring. We are driven to know more; to understand things that are beyond our own need to survive, or the need to propagate our species.

We want to know what we don't know; and as soon as we know what we don't know, we want to know that. Quite a conundrum!

RELIABLE KNOWLEDGE: THE ART OF PERSUASION

The goal of scientists employing scientific methods is to expand the body of Reliable Knowledge. Reliable knowledge is that knowledge society generally believes to be true, and therefore uses for societal purposes. Building bridges, growing food, and practicing medicine all rely on reliable knowledge. We base our designs, our practices, and our regulations on the body of reliable knowledge. But how reliable is "reliable"? Who is the arbiter of what is reliable, and what isn't? What is the ultimate test of the reliability of knowledge? Finally, how does Reliable Knowledge comport with a much broader view of Balanced Knowledge?

Ironically, the determination of Reliable Knowledge is not just a matter of applying Scientific Method; it in fact depends upon the ancient Art of Persuasion.

Scientists studying a problem conduct research, observation, testing, experimentation, and documentation. Based on this body of work, they develop assumptions, theories or postulates, and then conduct further study to verify these theories. Finally, Scientists must publish findings or scholarly works that demonstrate appropriate Scientific Method in developing and verifying such theories and postulates, in an effort to convince other equally qualified scientists of the "reliability" of the knowledge being espoused. Scientists must in fact persuade their peers of the validity of their work. Once a preponderance of peers within a scientific community accept the argument, the theory begins to be adopted as "reliable knowledge" by the general public; even though there may be serious disagreement or dissension within the same peer community.

A perfect example of the Art of Persuasion in determining Reliable Knowledge lies in the current issue of Global Warming. While there is growing evidence that the earth may in fact be getting warmer, there are still many in the scientific community who question whether the phenomenon is a long term trend, and many more that debate whether Global Warming is a result of Human interaction with natural systems on earth. Over the last several years, we have even seen indications that scientists may be doctoring data, or at the least, withholding or redacting information so as to better support a specific theory, one way or the other.

As a practical matter, polls demonstrate that most of the general public feels that the combination of a significant increase in the burning of fossil fuels coupled with a decrease in the world's forested land must have an effect on the accumulation of carbon dioxide in the world. Balanced Knowledge, or all Learning Experiences taken as a entirety, indicates that some environmental phenomenon is taking place that we refer to as Global Warming.

However, the cause and effect of Global Warming continue to be argued by scientists on both sides of the issue as competing interests politicize the debate in pursuit of different conclusions. So, here is a case where current observations flowing from all that we Know should be clear and compelling; and yet there is still disagreement over the Reliable Knowledge of Global Warming based on Scientific Method.

As you can see, the acceptance or rejection of the phenomenon of Global Warming will not boil down to merely objective information derived from scientific method. The issue of Global Warming will be resolved in the ability of the various constituencies to persuade the court of public opinion of the validity of evidence for or against the argument. Reliable Knowledge will be based on the most Persuasive Argument.

THE RELIABILITY OF "RELIABLE KNOWLEDGE"

Reliable Knowledge is also Fallible, and therefore not always Reliable. Reliable Knowledge is only reliable until it is disproved, refined, or replaced by other, more Reliable Knowledge. In fact, the body of Reliable Knowledge is constantly being refined, updated, redefined, or completely changed; which is why we must Learn continuously.

There are countless scientific theories that at the time of their acceptance, seemed to be strongly supported by observation and experimentation, and therefore became Reliable Knowledge; only to be subsequently debunked by new experimentation, observation, or reason. The earlier discussion on the development of the scientific principles of gravity demonstrates that yesterday's theories are replaced by today's understanding.

If you think such unreliable knowledge comes from past centuries; and that recent reliable knowledge is "more reliable", think again. We don't have to look back too far to see unreliable knowledge at work in our every day lives. Let's look at two 20th Century examples where Scientific Method provided "unreliable" Reliable Knowledge.

In medicine, the concept of Bloodletting a patient to cure a variety of ailments persisted from antiquity into the early 20th Century. It was even recommended in the 1923 edition of The Principals and Practices of Medicine, by Sir William Osler, the founding Professor of Medicine of Johns Hopkins University. By the mid-20th century, the thought of Bloodletting as a cure for anything had been completely discredited. Centuries of medical practices derived from reliable knowledge had been completely dispelled in a single generation.

An even more alarming and dangerous example of unreliable Reliable Knowledge based on science lies in the Theory of Eugenics. While I won't go into detail here (you can research Eugenics for more information), the Theory of Eugenics stems from the philosophy of Plato that government should control the advancement of the human race through genetic engineering. Ancient civilizations such as Rome and Sparta practiced infanticide to weed out the genetically inferior humans and build a stronger human race.

Sir Francis Galton, Charles Darwin's half-cousin, built upon his more famous relative's work by opining that man was not allowing Natural Selection to take its course among the human population. In fact, Galton theorized that by protecting the weakest members of society, man was weakening the entire human race.

Eugenics became a worldwide scientific movement in the early 20th Century as a U.S. scientist and biologist Charles Davenport established the Eugenics Record Office at Cold Spring Laboratory. Davenport's book *Heredity in Relation to Eugenics*, was used as a college textbook for many years. In the United States, eugenic sterilization became a popular state practice. In fact a report indicating favorable results of forced sterilizations in California was cited by the Nazi government to support Adolf Hitler's human experimentation in development of a master race; Nietzsche's theoretical Ubermensch.

Scientific study and development in Eugenics was conducted by prestigious universities, and funded by distinguished philanthropic organizations. Science had provided the Reliable Knowledge necessary to support societal acceptance of Eugenics, providing foundational support for anti-miscegenation laws, such as Virginia's Racial Integrity Act of 1924, which remained in force until 1967. Yet Science had not considered Balanced Knowledge in the concept of Eugenics, and the horrendous effect such a concept has on Human Interaction and the Human Condition.

As bad as the Theory of Eugenics appears to be, the Human Genome has once again awakened scientific wonderment regarding genetic engineering as a method of advancing the Human Condition. Clearly, such advances in our technological capabilities challenge us to answer a simple question. How Reliable is our Reliable Knowledge?

Balanced Knowledge vs. Reliable Knowledge

By now, it has become fairly clear that the term Reliable Knowledge is almost an oxymoron. Since all Knowledge is Fallible, even the most

Reliable Knowledge can be suspect. However, we still build things, discover cures, invent devices, construct structures; all based on our best available knowledge. Thousands of years ago, man was able to build incredible structures with significantly less knowledge than we have today. Today, we can cure diseases that a century ago we could not even diagnose. However, we also have higher incidence rates of a variety of diseases, from cancer to autism; which may in fact flow from our own advances.

An example of the unknown or unintended consequences of our Science can be seen in our food supply. In order to provide cheaper and more reliable quantities of meat protein, significant scientific advancements were made in the way we produced and processed beef in the United States, including the heavy use of hormones and antibiotics to allow for more densely populated feed lot production practices. Science improved production practices allowing us to deliver a relatively safe supply of high quality meat protein at a low cost. These advancements resulted in our ability to feed the growing appetite for beef and beef products within the U.S.

However, as increased levels of hormones and antibiotics entered the food supply, they began to have a direct and not altogether positive impact on human health. In addition, the higher level of animal fat in our diet resulted in increased rates of cardiovascular disease, as well as other gastrointestinal complications. Our short term scientific advancements in animal production practices have resulted in long term effects on a variety of Human Interactions, both good and bad.

The application of Scientific Method will continue to allow us to discover new "reliable knowledge" with regard to all aspects of our Universe, and our Life. Such discoveries are good, and continue to offer solutions to a host of challenges that confront the Human Condition. However, Human Behavior and Interaction should not be governed by what we Can do; they should be governed by what we Should do. Balanced Knowledge helps us discern between Can and Should.

Scientific Method cannot be the sole Learning Channel we use to determine how to react or respond to knowledge. We must constantly employ all three Learning Channels; Experimentation,

Tradition, and Reason, to advance our Learning and maintain Balanced Knowledge from which to base our actions. Experimentation and Reason must continue to push against the comfortable boundaries of Tradition; Tradition and Reason must temper the rapid changes offered by Experimentation; and Experimentation and Tradition must constantly test or validate what we Reason. Only when all three of these Learning Channels allow for a healthy exchange of thought can we acquire and maintain a level of Balanced Knowledge that will allow us to advance our understanding of the Universe and Human Interaction, and improve the Human Condition.

So it is with our scientific study of the various concepts of the Origin of the Universe and the Creation of Life. We need to evaluate all such theories critically, including the Theory of Supreme Intelligence; applying Experimentation, Reason, and Tradition. Only through Balanced Knowledge are we able to understand how Intelligent Forces can impact Human Behavior. Only then will we be able to address the complexities of Human Interactions and the resulting influence on the Human Condition and our World.

PART THREE: The Universal Stage

Chapter 6 - Everything is Something

As we enter the next Part of our study, we need to evaluate the Universal Stage including the sets and actors participating in the Universal Play. This discussion may seem at times elementary, and at other times tedious. However, in order to understand the Fundamental nature of Supreme Intelligence and Its relationship with the Fundamental Interactions of Matter, we need to understand the fundamental players on the Universal Stage. Therefore, please bear with me as we work through the setting for the rest of our story.

PHILOSOPHICAL PHYSICS

You have to love Physics. It's so logical, and yet can seem so abstract. Read a little about the dimensions of String Theory for some indication of where Physics can take us. So what exactly is Physics?

Physics, from Ancient Greek *physis*, meaning "nature", is a natural science that involves the study of matter and its motion through Spacetime along with related concepts such as energy and force. More broadly, it is the general analysis of Nature; conducted in order to understand how the universe and everything contained therein behaves. Pay particular attention to the last sentence. Basically, Physics is how we analyze Nature, our World, and our Universe. Sounds a bit like Philosophy, doesn't it? In fact, Physics finds its origin in Natural Philosophy.

Remember our discussion on Sir Isaac Newton? He published much of his work on Physics as the "Mathematical Principles of Natural Philosophy". As we follow the development of Physics, we see that it in fact emerges from the reasoning of Natural Philosophy, from the minds of Aristotle and Da Vinci, to Copernicus and Newton.

Even today, the philosophy of physicists will color their approach to understanding. Indeed it is very difficult to approach any subject without carrying our own Traditions (be they philosophical or not) with us. While Physics exposes knowledge through observation and experimentation, Philosophy exposes knowledge through reasoning and rational argument. In fact, many would say that the two cannot be separated, since reasoning and rational argument play an important role in experimentation.

As discussed in the previous chapter, both of these approaches play an important role in the three Learning Channels. Therefore, any discussion of physics starts with basic philosophical views - our Philosophical Physics.

SOMETHING: NOTHING ELSE MATTERS

Everything is Something unless it is Nothing. Another profound statement! From a physics perspective (staying away from abstract philosophy), we will start with this as another important Axiom.

Axiom 6.1: Everything is Something unless it is Nothing.

Based on this Axiom, we can evaluate everything in the Universe as one of two things - Something or Nothing.

Let's start with Nothing. Nothing is a little tricky, since it's hard to imagine NOTHING. Don't confuse air, or gasses, or invisible energy, with Nothing. Space, which is often considered Nothing, is in fact made up of both Nothing and Something. The Nothing of our Universe is not anything. It has no temperature, no heat, no light, no matter. It has no mass, no air, no gas, and no energy. It's stark, dark, emptiness – it's Nothing. The Universe may contain a lot of Nothing; since everything that isn't Something must be Nothing.

Why should we even look at Nothing? What difference does Nothing make?

The most important understanding we can take away from a discussion of Nothing is that Something CANNOT come from Nothing. This is the second Axiom we will advance. From our description of Nothing above it is clear that since Nothing is not anything, it cannot be the foundation of Something, initiate Something, or create Something. It is, after all, Nothing.

Axiom 6.2: Something cannot come from Nothing.

So, let's get right into our discussion of Something. If you think Nothing was tricky, wait until we try to get our heads around Something. We will begin with three fundamental and stable Somethings; **Matter**, **Mass**, and **Energy**.

Matter: The most visible and understandable part of Something is Matter. Matter is anything that occupies space and has mass. This is pretty familiar territory; basically, everything you can see, smell, or touch; anything with dimension; is Matter. In fact, even some things that you cannot see, smell, or touch, are Matter. It may exist as a solid, liquid, gas, or plasma. Matter is pretty much the Universe we are familiar with.

Matter is composed of particles. (WARNING - You may need a roadmap for this discussion!) Most of us are aware of molecules and atoms, those very small particles that comprise all Matter. However, those small particles can be further broken down into even smaller particles, such as protons, neutrons, and electrons. This starts to get to the limit of most of our learning. Yet we can take particle physics further by introducing the Mesons and Baryons; going on to the Fundamental Particles of Matter, the Quarks and Leptons.

Now we think we have arrived at the heart of the Matter. Or have we? These exceptionally small particles are considered Fundamental, since we do not believe they can be broken down further. However, we thought the same thing about the Atom; and then about Protons, Neutrons, and Electrons. The concept of Fundamental is important to this entire study, so while these particles seem to be too small to consider, their meaning is too large to ignore. What we have clearly determined is that there is more to matter than meets the eye.

We like matter. It's tangible. We can describe or define it; capture, collect, or control it; and most of all, own it! All the stuff we have; the things we possess; our collection of worldly treasure: that's all matter. In fact, sometimes it seems that the only thing that matters to us is Matter. However, as we all know, in the grand scheme of things, the matter we own, control, or possess, doesn't really matter.

Mass: We could not have Matter without the next Something; Mass. Mass adds "inertia" to our world and is often confused with Matter. But Mass matters more than Matter. All Matter contains Mass, but not all Mass is Matter.

Some will mistakenly use Mass when discussing weight, using the terms interchangeably. While they are related, they are not the same thing. Your mass is the same whether you're on the earth or on the moon, but your weight certainly varies.

Mass is a Fundamental property of an object that provides a measure of its inertia. Mass is the object's resistance to a change in velocity; its resistance to Force. Physicists spend careers trying to understand Mass. Witness the latest effort on confirmation of the Higgs Boson and the Standard Model, as scientists try to gain a better understanding of how Mass and Matter came to be.

Mass becomes even more complex as we attempt to understand it in relation to Special Relativity and the Mass-Energy Equivalence.

Energy: Which brings us to our third Something; Energy. Again, we can start with a simple defining phrase: Energy is the capacity of a system to do work. Energy is the "fuel" of the Universe. When released through time, it results in Power, driving everything that happens.

Potential energy is contained within a mass at rest. A cart sitting at the top of a hill has stored or potential energy as a result of gravity. Give the cart a push, and potential energy converts to kinetic energy, the energy of the object in motion. We might say that potential energy is standing by to do work, while kinetic energy is actually being consumed in the process of working.

Another form of Energy is that energy inherent in Matter through Mass. Based on the Theory of Relativity, the Energy of an object or system is equal to its Mass times the Speed of Light squared. This brings us back to the Mass-Energy Equivalence, and one of the most famous equations every written:

$$E = mc^2$$

We've tackled three pretty interesting Somethings; Matter, Mass, and Energy. We can see how they relate to each other; and how Matter could not exist without Mass and Energy. However, none of these three Somethings make anything happen. All of the Matter, Mass, and Energy in the Universe would simply vanish without three other equally important, but less understood Somethings: **Interactions**, **Force**, and **SpaceTime**.

Interactions: Let's start with Interactions. Without Interactions, nothing would exist. Mass would have no Mass without Interactions, and if you take away Mass, Matter and Energy must also disappear. If you recall we were introduced to Fundamental Interactions in a previous chapter, starting with Gravity. Gravity clearly exists, and is clearly Something. Yet it is not Matter, Mass, or Energy. It's an Interaction.

An Interaction is an action that occurs when two or more things have an effect on each other. Therefore, by definition, an Interaction cannot exist unless there are two or more things.

Interactions impact everything in our Universe and every activity in Life. Interactions are fundamental to biology, chemistry, physics, economics, genetics, communication, and on and on. Nothing could happen without Interaction. Interactions are required for the Universe to exist and for Time to Advance.

At the beginning of the Universe, we know that four Fundamental Interactions had to be established in very short order - nearly simultaneously. These Fundamental Interactions included Strong, Weak, Electromagnetism, and Gravity. Whether three of these Interactions emerged from a singular Unified Interaction, as in the Grand Unified Theory, is a question that is still being evaluated. However, we do know that all four Interactions came from Something. Based on our Axiom, they could not have come from Nothing.

Strong interactions hold the nucleus of atoms together, creating nearly all of the Mass. Electromagnetism keeps the electrons in their happy clouds spinning all around the atomic nucleus, allowing for the forces of electricity and magnetism, along with such things as light and microwaves. Weak interaction allows the quarks in our protons and neutrons to change flavors (yes, you read that correctly), allowing for radioactive decay. All three of those interactions hold our micro world together. Clearly, Matter, Mass, and Energy rely on Fundamental Interactions, which are a basis for what Physicists refer to as the Standard Model.

Finally, we must include the much loved but often misunderstood Interaction of Gravity to keep all the Mass in check once we get started. If we initiate the Universe without Gravity, all of the Matter, Mass, and Energy would simply fly apart. Nothing would coalesce; stars would not shine, galaxies would not form, planets would not exist or orbit.

Forces: Flowing directly from Interactions, we have Forces. Force is an amazing Something. Force moves us; allows us to overcome inertia; drives us forward in space and time. Forces are influences that cause things to change. Forces drive Behavior.

How do these Forces flow from Interactions? Once again we turn to Particle Physics for the likely answer. Remember the discussion of Fundamental Particles in Matter? Now we are going to consider Particles as Force Carriers. Under the current theory, each of our Fundamental Interactions is associated with a Force Carrier that is responsible for transmitting the Force. The alignment of Fundamental Interactions with Fundamental Particles includes:

1. Strong Interactions have Gluons
2. Weak Interactions have W & Z Bosons
3. Electromagnetic Interactions have Photons
4. Gravity has Gravitons (maybe; a really big maybe!)

Gravity, that pesky phenomenon! Science is not sure about the very hypothetical existence of a Force Carrier for Gravity; the Graviton. In fact, there are many who would argue that the concept of a Graviton creates even more difficult challenges. However, absent the Graviton, we have another problem – a Force without a Particle.

Could a Fundamental Force exist that is NOT associated with a Fundamental Particle? Well, we know that Gravity exists, and we know it interacts with Mass to create a Force. Yet the concept of a

Gravity Force Carrier Particle doesn't fit our observations; and is contrary to what we think we Know. Remember this problem because it will show up in a big way shortly!

So let's move from Force Carrier Particles back to Forces. Forces can change an object's physical geometry or state of being; modify its position or movement; alter its behavior or character; or change the object's energy state or attitude. We talk about Forces all the time; they impact every facet of our life. Yet, Forces remain mysterious.

Each of the fundamental interactions defined above result in Forces. Gravity interacts with mass and causes Force. Strong Interactions create exceptional powerful atomic Forces in the nucleus of atoms. Forces drive all actions in life. Without Force, nothing would ever change. Indeed, without Force, everything in the Universe would be static.

Forces also drive Behaviors; whether the behavior is the movement of inanimate tectonic plates; the competition of trees within a forest; or the struggle for human survival. Without Force, change would stop. Time would stand still.

SpaceTime: Finally we will examine what is probably the most difficult of our Somethings to comprehend, SpaceTime. The SpaceTime Continuum is the framework of Dimension. Without dimension, everything would occupy the exact same location at exactly the same time. No 3D, no 2D, not even 1D.

Everything; all of our Somethings, must exist in a dimensional framework. The framework we call our Universe is a continuum that consists of at least Four Dimensions; 3 spatial and 1 temporal. In SpaceTime, the three spatial dimensions of Length, Width, and Height, combine with the single temporal dimension of Time into a single mathematical model; the SpaceTime Continuum. The SpaceTime Continuum allows for observed phenomenon, and allows everything to occupy a position in space and time. Everything that has ever occurred since the Beginning of the Universe can be located in this four dimensional SpaceTime Continuum.

Before we move on, let's recap the six Somethings that we have established as fundamental pieces of our Universal Stage. They include **Matter, Mass, Energy, Interactions, Forces,** and **SpaceTime.**

Now we can get down to some serious discussion about one of our Questions: Where did all these Somethings come from?

Chapter 7 - Everything from Something

SOMETHING BEGETS SOMETHING

Everything that isn't Nothing came from Something. Therefore, in order for the Universe to Begin, the above Somethings had to come into existence; they had to be Created by Something.

Which Somethings existed at the Beginning of the Universe? Which Somethings were Fundamental to the development of the Universe? We can examine this by looking at the initial seconds of the Universe up to the point at which Matter began to form.

We will conduct this analysis in the context of the prevalent scientific theory for the Creation of the Universe, the Big Bang Theory. We will also couple this with our knowledge of the elements of the Standard Model, as well as the Grand Unified Theory and the Theory of Everything.

Something had to initiate the Fundamental Interactions that led to the Forces that drive Time and control Matter. Something had to create a Mass-Energy equivalence that contained all of the Mass and Energy for the entire Universe as we know it today. Something had to establish the SpaceTime Continuum, the dimensional framework in which the entire Universe exists. Finally, Something had to generate an incomprehensibly powerful Force that Triggered the atomic smashing of the original Mass-Energy particle that became our Universe.

All of these Somethings took place in a time period measured in yoctoseconds, or 10 to the minus 24th power. This complex initiation of each of these inherently different Somethings all came together at the same point all within less than a second. What caused all of these Somethings to occur with such precise timing, and work together in such a harmonious process? By definition, it had to be Something.

Therefore, we will add the following Axioms to our Balanced Knowledge.

> **Axiom 7.1: Everything came from Something.**
> **Axiom 7.2: The Universe was Created by Something.**
> **Axiom 7.3: The Big Bang was Triggered by Something.**

Chapter 8 - Forces and Equilibrium

MOTION CAUSES MATTER TO MATTER

We are all part of an extremely complex mind boggling Universe. Matter, the stuff that seems to matter to us the most, is in constant motion in SpaceTime. The motion may be the extremely large orbit of a solar system in a galaxy, or the extremely high speed oscillations of a subatomic particle. Matter is constantly in motion. The Universe is in motion. Life is in motion.

Because of Mass, Matter will not change its motion unless and until acted on by a Force. Unless a Force overcomes the inertia of Mass, the motion of Matter will stay constant. That would of course lead to a pretty dull Universe. No change, no action, no reaction.

The good news is that Forces are always in play, always doing Work. Forces are constantly pushing and pulling on Matter, resulting in a dynamic Universe that advances through SpaceTime. Work is defined as Forces moving Matter, or Mass, through a Distance. However, in the SpaceTime Continuum, Distance is not only length, width, and height; it is also Time. Therefore, it stands to reason that it takes Force to accelerate Mass through Time.

Despite this amazingly complex soup of diverse elements and matter being acted on regularly by tremendously powerful forces, the Universe, and more specifically our Solar System, is in a relatively stable state. This stability is not due to the lack of force, but is instead a result of a delicate balance of the various forces that drive the entire system. In fact, since the beginning of time, forces have never been Zero. Forces are always at work, making things happen, driving Mass forward in Time. Yet all these forces work in a wonderfully orchestrated harmonious balance. This condition of balance or stability is best described as "Equilibrium".

FORCES IN BALANCE

Equilibrium is a happy place. In a state of equilibrium, things Behave as anticipated. On a macro scale, our world is in balance. Life is good.

The reason Life is able to exist on Earth is because our Solar System, that small part of the Universe we inhabit, is in an amazing state of Equilibrium. No matter how chaotic and volatile we believe our personal lives to be, our solar system's behavior is very consistent; highly predictable. Likewise, Planet Earth is an equally stable platform for our habitation. For about 4.5 billion years Earth has spun around on its axis on a planetary orbit around our Sun; in a coordinated dance with other planets and other orbital and non-orbital objects that share this small piece of space. No random bouncing around of planets here.

The Earth circles the Sun in a precise orbit and with split second timing. Likewise, our Moon circles the mother ship with equal precision and accuracy. Focusing a little closer to home, we find that Behavior within the Earth's atmosphere is also quite predictable. Tides are scheduled to the minute, and ocean currents follow the same paths for centuries. Over time, even weather and the movement of continents follow patterns that are certainly not random; but instead part of this phenomenal balance of Forces - Equilibrium.

As with the macro perspective, the microscopic view of our world is similarly fascinating; similarly stable. Atomic structures composed of highly energized subatomic particles bouncing around within their structures are at equilibrium when joining with other atoms in the variety of molecules that comprise all of matter. Just as the Universe around us, the Universe within us stays in equilibrium through an astonishing balance of very strong forces that are yet to be fully understood.

Equilibrium, however, is not static. As stated previously, the Universe, macro and micro, is constantly in motion. From the expanding Universe to our individual daily lives, we are part of a very dynamic system. Things move; Behaviors change; the Earth revolves. In other words, Life "oscillates".

In every day parlance, these oscillations are referred to as the ups and downs of living. We live in a constant state of undulation, as the forces that surround us push and pull on our physical and social environments. These varying forces influence Behaviors, which in turn cause new Interactions, in a never ending cycle. Earth's orbit around the sun changes as the Earth pivots on its axis. These oscillations are however consistent; the undulating movement precisely predictable.

Likewise, our Moon's orbital pattern changes consistently. The effect of these two oscillating bodies creates a force synergy that drives tides, changes weather, and provides an environment for life. We can't stop these cycles any more than we can stop Time. Indeed, we wouldn't want to, since Life and Time would also stop.

The Ultimate Goal. The ultimate goal in understanding the forces that drive our Universe and Life is our ability to apply Human Intelligence to achieve appropriate Equilibrium. We can use Intelligence to mediate the Forces that drive Human Behavior, which in turn are instrumental to an appropriate balance in social systems and structures. We seek to find a state in which the oscillations of Human Interaction are relatively steady. Behaviors can be anticipated and events occur as expected. This is not only a matter of Human survival; it's a matter of our World's survival.

In the last one hundred years we have seen how Human Intelligence can throw the Balance of Natural Forces out of Equilibrium. We can use the Forces of Human Intelligence to destroy the Earth through the reckless exploitation of nuclear fission or the negligent actions that lead to Global Warming. Or we can use the same power of Intelligence, governed by the fundamental Nature of Supreme Intelligence, to balance Human Interaction and solve the myriad challenges impacting Life and our World.

The key to achieving a balance in the Forces that drive Human Behavior and a corresponding Equilibrium in Human Interaction is the recognition of the Theory of Supreme Intelligence.

PART FOUR: The G-PHENOM

Chapter 9 - The Universal Phenom

THE MASTER TRIGGER

In The Beginning. Imagine if you will, a Universal Phenomenon that originates four Fundamental Interactions generating the Forces controlling the entirety of Matter in the Universe; Forces that shape heaven and earth; move mountains; part seas (Does this sound familiar?)

This Phenomenon also created all of the Mass and Energy for the entire Universe for the rest of time, and packed it into an infinitesimally small particle of almost infinite density. It initiated a framework that provides the spatial and temporal dimensions allowing everything we know to exist. This Universal Phenomenon initiated the Big Bang; produced light from absolute darkness; started the ticking of Time. This inconceivably powerful Phenomenon propelled the entire Mass of the Universe along the Spacetime continuum at unimaginable speeds, for the rest of Time.

This Universal Phenomenon provides for the Creation of the Universe as a single coordinated event caused by a single master actuator, a Master Trigger. This single Master Trigger initiated a series of interdependent events, including the development of the four Interactions, the creation of a Mass-Energy Equivalence, and the establishment of the SpaceTime Continuum. The single Master Trigger also created the unimaginably powerful force required to trigger the Big Bang.

Based on such a Universal Phenomenon, we can create a flow diagram of the Creation of the Universe. We can map out and analyze the series of specific dependent activities using a critical path analysis, with all dependent activities flowing from a single Master Triggering Event. This single Universal Phenomenon, or Master Trigger, parallels the combination of two other Scientific Theories, the Grand Unified Theory and the Theory of Everything.

GUT, an acronym coined by CERN Researchers in 1978, refers to The Grand Unified Theory that hypothesizes that the three gauge interactions of the Standard Model (Electromagnetic, Strong, and Weak) may have been derived from a single combined Interaction. Taken a step further, if we merge the single Unified Interaction of the GUT with the Interaction of Gravity through another unifying Phenomenon, we arrive at the Theory of Everything ("TOE").

A single Master Trigger would be independent of any other entity, Force, or Phenomenon, and therefore would truly be Fundamental. A single Master Trigger hypothesis for the Big Bang is a significantly stronger model than the alternative concept requiring the spontaneous, accidental, random occurrence of four unrelated events at precisely the same time that the Mass-Energy Equivalence and SpaceTime Continuum are ready to go.

A single Master Trigger accounts for the coordinated creation of the four Fundamental Interactions of Matter and initiation of the Big Bang; allowing Mass and Energy to exist and SpaceTime to begin. At this point in our study, we can call this single initiating Phenomenon anything we like. I will take an author's privilege in this regard, and will designate this single Universal Phenomenon as "G"; or the G-Phenom.

Therefore the following Postulate is offered.

> **Postulate: The G-Phenom was at The Beginning of the Universe, created the elements required to Initiate the Universe, and Triggered the Big Bang.**

THE ANTI-HYPOTHESIS

The strength of the G-Phenom Postulate lies not only from the evidence stated above, but also from the lack of any reasonable or credible Anti-Hypothesis. Let's examine several possible challenges as a starting point for the bombardment of criticisms that should flow from such a Postulate.

By Any Other Name: The first attack will probably be an emotional one. In this case, the knee jerk reaction might be "It can't be G!" It must be something else.

Some other Thing must have pulled the Trigger of the Universe starting gun. To that argument, I would merely respond that you could call it whatever you want. At this point, I have not ascribed any Characteristics or Behaviors to the G-Phenom, other than (1) It was at the Beginning, (2) It was Fundamental, and therefore came from no other source, and (3) It served as the Trigger, or Creator of the things that flowed from It. These three Behaviors align at least in part with the concept advanced by the Theory of Everything.

So, we could call it the A-Phenom, the E-Phenom, the Y-Phenom, or even Chloe the Phenom (named for my Dog). Simply renaming it does not change the Postulate.

We Can't Know: The second attack is a familiar one. Many very learned readers will simply state that the G-Phenom doesn't exist, or can't exist, because we don't Know IT. We can't Prove It.

This argument has become very popular in the last century with regard to theories regarding God. Some in the scientific community have adamantly denied the existence of such a Phenom, not as the result of an equivalent alternate theory, but simply as a result of their unwillingness to accept the existence of such a Phenom. This criticism simply holds no water.

An unwillingness to accept a logically presented Postulate that flows from other widely accepted Axioms, does not disprove or even shake the Postulate. It merely calls into question the capacity or motive of the criticism. Science currently infers that most of the Matter and Energy of the Universe is Dark Matter or Dark Energy, with no other basis than the fact that it must be there or our other knowledge is wrong. Science hypothesizes a Graviton so they have a Force Carrier for Gravity, despite the fact that it does not comport with what we observe. Clearly we can and do form scientific hypotheses, theories, and postulates based on much less evidence than that demonstrating the existence of the G-Phenom.

It Isn't Science: Hopefully we put this argument to rest in Chapter 1. This is a Postulate based on the best Science we currently have. It parallels other scientific theories currently being evaluated, such as GUT and TOE. If you want to read more about this Science, research the Grand Unified Theory and the Theory of Everything. These theories align closely with the Phenomenon of a Single Master Triggering Event; the theory which leads to the Postulate of the G-Phenom.

In fact, it is difficult to understand how we can allow for the scientific discussion of GUT and TOE without ultimately allowing for scientific study of the Postulate of the G-Phenom. The only reason to argue against the G-Phenom is to attempt to avoid the inevitable discussion that will naturally follow the Postulate itself: that GOD IS A Matter of Fact.

The above are certainly not all the criticisms, difficulties, or opposing theories that may be proffered. However, any alternate theory would have to address the simultaneous creation of four Interactions, the creation of the Mass-Energy Equivalence (no matter how small or large the particle containing the entire Mass might be), and the creation of the SpaceTime Continuum. These ingredients had to all be available at the Universe's starting point, and they all had to be Triggered by Something - a Master Triggering Phenomenon.

Which brings us right back to our Postulate:

> **The G-Phenom was at The Beginning of the Universe, created the elements required to Initiate the Universe, and Triggered the Big Bang.**

Chapter 10 - What is the G-Phenom?

REPRISE: EVERYTHING IS SOMETHING

Remember our previous Axiom? Everything is Something; and the G-Phenom is no exception. It certainly is not Nothing.

But what is it? Could it be one of the fundamental Somethings previously discussed? It is not likely we will be satisfied leaving this question open, so let's see if we can give additional definition to the G-Phenom. We will start by comparing the G-Phenom to the six previously defined fundamental "Somethings"; including Matter, Mass, Energy, Interactions, SpaceTime, and Force.

Starting with Matter, we will find that we can immediately remove this Something from further consideration, since it is not really "Fundamental". While Matter seems to matter the most to most of us, Matter wasn't there from the Beginning. Matter is dependent upon the other Fundamental Somethings for its existence. Based on what we know of the Big Bang at this time, Matter was probably not even introduced to the SpaceTime Continuum until minutes after the Trigger, when the atomic nuclei of hydrogen and helium began to form, and atomic structure started to develop. This is a long duration in the yoctosecond timing of the Big Bang. So, Matter will no longer matter in the G-Phenom definition.

Next up we will examine the G-Phenom as a Mass-Energy equivalence. Based on the Theory of Relativity, Energy cannot exist without Mass, and vice versa. So in the context of the G-Phenom, we cannot look at them independently. We also know that Mass and Energy are dependent upon the existence of the Interactions of the Standard Model, and could not have preceded those Interactions. Since the G-Phenom initiated the Fundamental Interactions, and the

Fundamental Interactions preceded the Mass-Energy equivalence, the G-Phenom cannot be a Mass-Energy Equivalence.

Three Somethings down, Three Somethings to go.

The Multi-Dimensional SpaceTime Framework. Let's consider the G-Phenom in the context of a multi-dimensional framework, such as the SpaceTime Continuum. In other words, maybe the G-Phenom is a multi-dimensional container in which everything exists; and from which everything is derived. This thought has some possibilities. After all, the entire Universe exists in the SpaceTime Continuum.

Could emerging multi-dimensional String Theories be the key to the Beginning of the entire Universe? Such concepts do in fact contend for consideration in the Theory of Everything. Superstring theories emerging over the last several decades require more than the four dimensions currently observed in the SpaceTime Continuum. Most recently, the M Theory supposes the existence of an eleven dimension framework for Everything in the Universe. There are two paradoxes to the concept of such a multi-dimensional framework serving as the single fundamental Triggering Phenomenon for the Beginning of the Universe.

First, String Theory requires subatomic Force Carrier particles (bosons, photons, gravitons) and they exist in oscillating motion. String motion therefore requires energy; potential or kinetic. Think of it as a guitar string under tension. Strum the string and it will vibrate. However the oscillating string will not continue to oscillate without one of two things - an interaction (getting strummed again), or kinetic energy that will allow it to just keep vibrating for as long as it has energy. Since string theory relies on the existence of numerous Fundamental Particles as well as Energy, we are back to the same dilemma encountered with regard to the Mass-Energy equivalence. How could several of the Somethings that were created by the G-Phenom as part of a larger Trigger Phenomenon, be the G-Phenom?

Second, String Theory is a mathematical model that mediates the differences in classic physics, relativistic theory, and quantum mechanics. The Theory builds on complexity instead of breaking down to simplicity.

String Theory attempts to describe all fundamental Forces and Matter. However, in doing so, it introduces a number of additional unobserved hypothetical fundamentals that must also exist in order to reconcile the three schools of thought it was attempting to mediate. For example, String Theory has been divided into five different supersymmetry strings involving ten dimensions. The Membrane Theory, one of the most recent iterations of String Theory, purportedly reconciles the five superstring theories providing eleven dimensions and introducing the "p-branes" and brane cosmology. (Do NOT confuse "p-branes" with pea brains).

However, the basic for the Membrane theory (membranes) dropped out of this Theory as it became the M-Theory, which became the unifying theory of the Superstring Theory. This sounds surprisingly like what the Grand Unified Theory attempts to do for the Gauge Interactions. As with GUT, M-Theory leaves us with same problem Interaction - Gravity.

String Theory reconciles this problem by hypothesizing the Graviton; which as we have previously discussed, is a major problem in itself. So, it appears that the complexity of String Theory, the introduction of additional dimensions, and the creation of new hypothetical Somethings in order to make it all work actually takes us further away from a singular Something under the Theory of Everything.

Therefore, the SpaceTime Continuum and String Theory do not appear to represent a viable explanation for the clear, concise, and pragmatic Postulate of the G-Phenom.

A Master Interaction. Since Gravity and the Super Interaction of GUT seem to consistently challenge our other theories, we should examine the potential that the G-Phenom is the combination of these two Fundamental Interactions. Could the G-Phenom be a single Master Triggering Interaction that spawned Gravity and the GUT Super Interaction, and then executed the Big Bang. This concept is at first appealing in its simplicity; and in its alignment with GUT and TOE. Unfortunately, we have one major definitional problem with this hypothesis; the term Interaction.

As you will recall, an Interaction is an action that occurs when two or more objects have an effect on each other. Since the G-Phenom is a Singular Phenom that existed prior to the Beginning of the Universe, and nothing else existed at that point, it cannot possibly fit the definition of an Interaction.

The G-Phenom clearly "acted"; it initiated all of the requirements of the Universe and Triggered the Big Bang. However, the G-Phenom could not have interacted with anything else, since there was nothing else with which to interact. The G-Phenom must have acted on its own, without the Interaction of any other Something. So, we can strike Interaction off the list.

Force. We are now down to one Something that we can use to help us define the G-Phenom - a Force. Looking back to our initial discussion on Something, we recall that Forces are influences that cause things to change. Forces alter the momentum of a system. Absent a Force, momentum cannot change.

This seems to fit our Postulate fairly well. The G-Phenom clearly influenced the Universe, since it both initiated all of the requirements and triggered its existence. That is a significant influence. The G-Phenom also clearly caused a change. Before the Big Bang, no Universe. After the Big Bang, a Universe. Voila, Change!

Therefore, looking at the Postulate of the G-Phenom in the context of our definition of Force, it is apparent that the G-Phenom at the very least exerted a substantial and significant Force. However, this simply brings us back to our original Postulate wherein the G-Phenom exerted the Force that Triggered the Big Bang.

As with our other Somethings, we quickly find a problem describing the G-Phenom as simply a Force. A Force cannot trigger itself. A Force is a response to an Interaction, such as gravity and electromagnetic. All Forces pertaining to Matter evolve from the Fundamental Interactions and can be derived from those. Forces can flow from chemical interactions, biological interactions, and nuclear interactions; to name a few. However, in the case of the G-Phenom, there was nothing to interact with.

The G-Phenom was Itself Singular and Fundamental; therefore It had to initiate the Force. While we can certainly ascribe the initial

Force triggering the Big Bang to the action of the G-Phenom, we cannot limit the G-Phenom to only the Force. By creating the Mass-Energy equivalence, initiating the SpaceTime Continuum, and Triggering the Big Bang, the G-Phenom demonstrated behavior beyond that attributable to force alone.

TO SOMETHING AND BEYOND

At this point, our effort to further define the G-Phenom has not progressed very far. This is probably because we tried to use our list of the Somethings created by or initiated by the G-Phenom to describe the G-Phenom.

We shall therefore take another approach in our quest for the Essence of the G-Phenom. Instead of looking at the results of what the G-Phenom did, we shall examine the actions of the G-Phenom as an indication of its Behavior. These actions can be written out as a process diagram for the Beginning of the Universe. Such a process diagram may include the following steps:

1. The G-Phenom initiated the fundamental Interactions, more than likely as a single Super Interaction (GUT) coupled with the independent Interaction of Gravity.

2. The G-Phenom established a SpaceTime Continuum as a framework for the initiation of the Universe, providing spatial and temporal dimensions for the Universe to expand into.

3. The G-Phenom created a Mass-Energy Equivalence containing all of the Mass and Energy for the entire Universe for the rest of time.

4. The G-Phenom Triggered the Big Bang with a Force that exploded the Mass-Energy Equivalence and accelerated it into the SpaceTime Continuum.

One might argue the order of the above steps, or the dependence of one or more upon the other. For example, did the Interactions of Step 1 have to exist to support the Mass-Energy Equivalence of Step 3, and therefore have to precede it? Whatever the order of the first 3, it is clear that they had to occur for Step 4, the Big Bang to be triggered.

Otherwise, the Universe could not have formed. We can certainly study the arrangement and critical path of these steps – but all four steps had to exist for the Expansion of the Universe to begin.

The Big Bang process described above allows us to better understand the behavior of the G-Phenom. The G-Phenom didn't actuate just one thing; it actuated a series of distinctive steps in a very precise yoctosecond timeframe. From these actions, we can surmise that the G-Phenom was able to act upon an environment; determine discrete actions required to achieve a goal; and then employ actuators to cause the actions and achieve that goal. This complex set of actuations clearly demonstrates at a minimum that the G-Phenom exhibits the characteristics we define as Intelligence.

We cannot know if the G-Phenom applied critical thinking, reasoning, abstract thought, or problem solving when Triggering the Big Bang. We simply don't have evidence of what took place before the beginning of the Universe to know what the G-Phenom went through in deciding to create the Universe. That, however, does not prevent us from evaluating the actions that can be attributed to the G-Phenom, and from that establishing a behavior. At a minimum these actions are the defined Behavior of Intelligence.

Could the G-Phenom be a form of Intelligence? Its Behavior certainly indicates that it could be. Could the G-Phenom have been some form of Antonymous Intelligent Agent? Intelligent Agents must derive their intelligence from another intelligent source. They cannot simply spawn themselves. Intelligent Agents must act as Agents of an Intelligent Being, an Artificial Intelligence, or another Intelligent Agent. The G-Phenom by definition was in the beginning, and did not derive anything from any other source. It was the original Something. It could not have been an Intelligent Agent of another Intelligent Being or of a created Artificial Intelligence; since nothing else existed. Therefore, the G-Phenom must have initiated its own Intelligent Behavior. We can further surmise that the G-Phenom's actions indicate Innate Intelligent Behavior.

Therefore, we can refine our Postulate as follows:

Postulate: The G-Phenom is the Singular Fundamental Intelligent Phenomenon that Initiated the Creation of the Universe and Triggered the Big Bang.

DETERMINING CHARACTER FROM BEHAVIOR

Characterizing the G-Phenom as Intelligent seems to be a giant leap of "faith"; but this characterization has nothing to do with faith, tradition, or revelation. The observable evidence that allows us to make such a statement has been addressed. A classic way to determine the character of something is to analyze the behavior of that thing. We used such Behavior analysis in order to ascribe the characteristic of Intelligence to the Phenom.

We often describe the characteristics of things based on their behavior without regard to their motive. So it makes no difference why the G-Phenom acted. Therefore, to critically evaluate the above refinement of the Postulate, we should look at the alternatives to the recitation of the actions (the "**Behavior**"), and other possible descriptions we can assign to this behavior (the "**Character**").

Behavior. For now, we have associated the G-Phenom with a single objective: creating the Universe through the Big Bang. This objective required a series of fundamental things to occur in order to complete the action - we will call these things the Behavior. At this point we are not examining what has happened since the Big Bang; that will be taken up later. We are instead looking at what took place in order to affect the Big Bang, which is what the Postulate of the G-Phenom currently addresses.

The Big Bang Theory itself says a lot about what led up to it. The Big Bang Theory of the Universe is the opposite of the "slow fizz" Theory of Evolution. Our best scientific knowledge demonstrates that the Universe did NOT evolve slowly. It did NOT begin as a single particle; a boson, quark, photon, or graviton; and slowly grow from there. It did not start out as a random singular effect, such as Gravity, and begin to spawn itself. The Universe started in a really BIG BANG. This is a major consideration in the development of the Postulate.

The Big Bang was almost instantaneous, the series of actions taking place in the realm of the yoctosecond. Yet, it required more than a single activity; or a single interaction between two phenomena. A series of separate fundamental activities became steps in a Behavior that was required to achieve the objective of the Big Bang. A simple list of these fundamental activities is shown in the four steps discussed previously. So, for our first critical analysis we need to look at the four actions that make up the G-Phenom's Behavior, and see if any of them could be eliminated as requirements to achieve the objective.

For example, in order to effectuate the Big Bang, did the Four Fundamental Interactions need to exist; or at the very least two; Gravity and the GUT Super Interaction? How about the SpaceTime Continuum? Did we need a multi-dimensional framework into which the Universe could expand and Time could begin? Was a Mass-Energy Equivalence necessary at the time of the Big Bang? If not, what Banged? Finally, was a Triggering Force necessary to overcome the initial momentum of the Mass Energy Equivalence and start the expansion of the Universe?

These are the actions that indicate the Behavior of the G-Phenom. While they can and should be studied, analyzed, and debated; at this time, it is impossible to eliminate any of them as requirements to meet the objective; Triggering the Big Bang and Creating the Universe. Therefore the four actions described above are a reasonable approximation of the G-Phenom's Behavior.

Character. The next step is to classify, or characterize, this Behavior. What Characteristic can we apply to Something that Behaves as the G-Phenom did? Above, we classified this Behavior as Intelligent, and therefore ascribed this Characteristic to the G-Phenom. The opposing hypothesis to characterizing the G-Phenom as Intelligent would be to describe the G-Phenom as Un-Intelligent. This would require the Behavior of the G-Phenom to be a result of a random accident that happened to initiate all the actions and trigger the Big Bang. We need to evaluate this hypothesis also, since our Postulate must choose one or the other Character - either the G-Phenom is Intelligent or it's Un-Intelligent. So, let's examine the alternate hypothesis and try to resolve the dilemma.

Counter Hypothesis: An Un-Intelligent Phenom. As stated above, we don't know what the G-Phenom did prior to the Big Bang, or what process it might have used in arriving at a decision regarding the complex variables that resulted in the Big Bang. We cannot at this time ascribe a motive to the G-Phenom.

However, just as with our analysis above, we do have a great deal of clarity into the elements that were required to initiate the Universe and Trigger the Big Bang. Therefore, let us see how an Un-Intelligent Phenom would have caused those elements to come into existence.

Absent any intelligence, the G-Phenom's actions would have had to be random, accidental, and spontaneous ("RASP"). Otherwise, they would have been planned, purposeful, or deliberate - which would demonstrate Intelligence.

The G-Phenom's first action would have been the RASP initiation of at least two Fundamental Interactions, Gravity and the GUT Super Interaction. Simultaneously, the G-Phenom would have initiated a RASP dimensional framework in which these interactions and "coming attractions" could exist and move or advance - a SpaceTime Continuum. Next up, the G-Phenom must create a RASP Mass-Energy Equivalence containing all of the mass and energy required to produce all the Matter in the upcoming Universe. This Mass Energy Equivalence must be held in a super dense super small state by the GUT Super Interaction, since nothing else has been created or initiated.

The G-Phenom must now initiate a fourth and final RASP action and trigger a counter force to the Super Interaction; a particle smashing force at least as strong as the entire Mass Energy Equivalence of the Universe in order to trigger the Big Bang and blow apart the very Particle that was just randomly and accidentally created.

This RASP Force also had to breakdown the GUT Super Interaction into three new Fundamental Interactions; Electromagnetic, Strong, and Weak. These new Fundamental Interactions, along with Gravity, must exist in order to allow the rapidly expanding exploding sub-atomic particles to coalesce into nuclei, electron clouds, atoms, and finally molecules; creating the foundation for Matter.

The simultaneous coupling of any two of these complex actions as RASP events is highly improbable. The RASP eruption of all four actions in a specific order in a period of less than a second is a statistical impossibility. As an Un-Intelligent phenomenon, the G-Phenom would have had to randomly, accidentally, and spontaneously cause these actions with no planning, design, deliberation, intention, or purpose.

Both hypotheses are now on the table. The G-Phenom's Behavior must be characterized as either Intelligent or Un-Intelligent. There is not a third choice.

So what do you think?

Based on the definition of the characteristics of Intelligent behavior; the lack of any other interaction, cause, or intelligent influence for the G-Phenom's Behavior; and the statistical improbability that all of the actions that indicate the Behavior were random, accidental, and spontaneous; it becomes apparent that the evidence strongly supports the hypothesis that the G-Phenom can be Characterized as Intelligent based on Behavior. The G-Phenom's actions are consistent with Intelligent Behavior.

Therefore, the refined Postulate stands:

The G-Phenom is the Singular Fundamental Intelligent Phenomenon that Initiated the Creation of the Universe and Triggered the Big Bang.

Chapter 11 - An Intelligent Being

TO BE OR NOT TO BE

Through an analysis of Behavior, we were able to make some progress on trying to understand the G-Phenom. However, we are still not able to get our hands around what the G-Phenom is, or how we might better describe it other than a Phenomenon; an observation. We have ruled out the complete current library of Somethings. By definition the G-Phenom cannot be Nothing. So, let's look at a few of the things we do know and see if we can take another step forward in our quest to define the G-Phenom through application of some additional scientific study.

> 1) The G-Phenom is a Singular Fundamental
> Phenomenon.
> 2) The G-Phenom demonstrated Behavior consistent
> with and indicative of Intelligence.
> 3) The G-Phenom initiated a Force.

We have certainly deliberated items 1 and 2 at length, which has gotten us this far in our effort. Item number 3 however provides a new avenue for potential discovery. Let's examine it further.

As we know the G-Phenom had to initiate a force in order to Trigger the Big Bang. Forces are not self-initiating. A force cannot spontaneously occur on its own accord. We should therefore consider the things that can initiate a force, or cause a force to happen. This should provide additional insight into the G-Phenom.

Generally, all Forces in the Universe are triggered by Interactions. As you recall, an Interaction is an action that occurs when two or more things have an effect on each other. The Standard Model defines three interactions as Fundamental Interactions; and assigns Force Carrier particles to each. Gravity of course is the fourth

Fundamental Interaction; and we have already discussed the difficulties encountered with the hypothesis of the Graviton. These four Fundamental Interactions were required at the beginning of the Universe, and initiate all the Forces required for Mass and Matter.

We regularly observe the forces resulting from all Four Interactions. The Forces resulting from Gravity and Electromagnet Interactions are obvious to casual or normal observation. The Force effects of Strong and Weak Interactions, although not as readily observable, are equally as important. Forces also result from the derivatives of these Fundamental Interactions. Chemical reactions result from molecular level interactions initiating certain forces. Nuclear interactions and biological interactions result in forces. Nearly all forces can be directly attributed to Interactions. Based on the above, the following two Axioms are offered.

Axiom 11.1: Forces are not Self Initiating

Axiom 11.2: Interactions initiate Nearly All Forces

Why does Axiom 11.2, state "Nearly" all Forces? Why not simply say All Forces? The Answer lies in our development of the Postulate of the G-Phenom.

You may recall the following progression of thought. There is strong scientific evidence that indicates that the Universe was initiated by a Singularity. This is the basis of the Theory of Everything. In addition, it is clear that a force was required to Trigger the Big Bang. Therefore, that Force must have been initiated by the Singularity. By definition, the Singularity cannot be an Interaction, since there are not two singularities to Interact. Therefore, something other than an Interaction Triggered the Force leading to the Big Bang.

This inevitably leads to the conclusion that there must be another Something that can initiate a Force. If we call this other Force initiating Something the G-Phenom we have just used circular logic to go nowhere. We have defined the G-Phenom as the G-Phenom; and therefore we failed to advance our understanding at all.

We must instead present a new Theory to further define the G-Phenom; then test the Theory for validity. For this exercise, we will start with the three statements presented above; (1) The G-Phenom was a Singular Fundamental Phenomenon; (2) The G-Phenom demonstrated Behavior consistent with and indicative of Intelligence, and (3) The G-Phenom initiated a Force.

From these statements, we will consider the following Theory:

Theory 11.1: Theory of Intelligent Forces: Forces can be Initiated by the Intelligence of Intelligent Beings

As we have done previously when we introduce a hypothesis or theory, we must evaluate this new statement to determine if it is well supported. Many will accept the Theory of Intelligent Forces as an Axiom based on their personal observations of Intelligent Beings. However, for those who either question or simply don't believe this statement, we must provide clear evidence supporting the Theory. For this discussion, we will focus our reasoning and conversation on the observations of the largest pool of the most Intelligent Beings that we are currently aware of in our Universe - Humans.

First we should clear the issue of whether a singular inanimate object can initiate a force. There is no evidence that a rock or a lump of coal can singularly and spontaneously initiate a force. Even a radioactive substance cannot initiate a force absent the Fundamental Interactions. Inanimate objects have no signaling or self-initiating capability. They are completely and totally reliant upon the Fundamental Interactions for all of their Behavior. In fact we can take this a step further by recognizing that inanimate objects are not even Singularities. Every inanimate object is a complex piece of matter that is constantly in a state of Interaction within itself. This can be said for all Matter, animate or inanimate. Matter would not exist at all without the Fundamental Interactions.

This does not however negate our Theory. While matter exists as a result of the forces created by Interactions, matter cannot self initiate any of those forces. Matter simply exists - Interactions trigger the forces. This observation is entirely consistent with the Axioms presented thus far.

The Theory of Intelligent Forces does not rely upon the Interactions of Matter. Instead, it proposes that forces can be initiated by a Fifth Independent Fundamental Phenomenon; a Phenomenon we can define as Fundamental Intelligence. The Theory of Intelligent Forces recognizes that Intelligence must flow from an Intelligent Being. By definition, Intelligence cannot flow from an unintelligent entity.

Our scientific study of Intelligent Forces must evaluate through reason and observation the Forces that are initiated by Intelligent Beings. If we find that Humans can independently initiate Forces through Intelligence, then we can state with certainty that Intelligent Beings can initiate Intelligent Forces. The ability of an Intelligent Being to independently Initiate Forces would explain the Behaviors exhibited by the G-Phenom, and offer a plausible solution for the Theory of Everything.

Self-Initiating Behavior. Intelligent Beings appear to demonstrate Self-Initiated Behavior. Put another way, Intelligent Beings appear to be able to make individual choices or decisions. The question is whether such Self-Initiated Behavior is solely the result of the four Fundamental Interactions of Matter; or whether another independent Fundamental Phenomenon is at work - Intelligence. Observation of our pool of Intelligent Beings demonstrates that the answer is that Forces can flow from either Intelligence of Interactions; depending on the type of resulting Force.

There will be those who argue that all the forces generated by Intelligent Beings are in fact simply a result of Interactions within the Matter that composes the Being. In other words, they would theorize that every force that Intelligent Beings initiate is a result of one of the Fundamental Interactions triggering the biological functions within the person or animal.

We must recognize that the Fundamental Interactions of Matter do initiate a variety of biological forces. The basic life functions that are observable across all living organisms are for the most part controlled by the Fundamental Interactions of the Matter of the living organism. A vast array of biological, chemical, electrical, mechanical, and other cellular, molecular, or atomic activities within a living organism are derived from the four Fundamental Interactions. Involuntary functions (heart beat, breathing, digestion, etc.) and reflexive functions are examples of such Interaction based Forces.

However, there are a number of observed Behaviors in our Human pool that do not appear to flow solely from the Fundamental Interactions of Matter. In fact, it appears that Intelligent Beings exhibit Behaviors that are contradictory to what we might expect if all Forces were derived from automatic unvarying Interactions.

The Big Bang required a force. That force could not have been triggered by the Interactions that were created as part of the Big Bang! Therefore, there must be another Source of the Force demonstrated by the G-Phenom in initiating the Big Bang and Creating the Universe. The Theory of Intelligent Forces and the Theory of Supreme Intelligence respond elegantly to this problem.

FUNDAMENTAL INTELLIGENCE

The introduction of Fundamental Intelligence flowing from a Supreme Intelligence creates an entirely new dynamic in our discussion of the Fundamental Phenomenon of the Universe and the Laws of Nature. Our observations indicate that Intelligence allows an Intelligent Being to make independent decisions and choices. This is directly contrary to the concept of Fundamental Interactions, which do not allow for the concept of choice.

Fundamental Interactions behave exactly the same, all the time, everywhere. Even when observations indicate a difference, these differences can be reconciled so that the Interaction behaves consistently. Our previous example of Gravity clearly demonstrates such a reconciliation of observation. Our friends living in space or under the ocean observe gravity differently than we do; but the Interaction is actually behaving consistently in all realms.

Intelligence, however, breaks the certainty of this model. The Forces that drive the behavior we describe as Intelligent cannot be attributable to the four Fundamental Interactions. Again there may be those who argue that everything we call Intelligence is a matter of simple brain chemistry, and that making Intelligent decisions is merely the result of the four Fundamental Interactions working within cranial gray matter. This is the concept of Matter over Mind. This argument would contend that all of the actions of an Intelligent

Being are driven by the Matter of the brain alone. However Matter over Mind falls apart in light of the actual observances of Intelligent Beings.

Attributing all Forces of Intelligent Behavior solely to the strict governance of four Fundamental Interactions in Matter leaves several major quandaries.

1) It would completely eradicate the concept of Intelligence. If the four Fundamental Interactions rule all Intelligent behavior, and by definition the four Fundamental Interactions always behave consistently, then they would not allow for things such as critical thinking, abstract thought, or reasoning. Intelligence could not rule Interaction; Interaction would rule all Intelligence. There would be no impetus for the development of Intelligence, since all of the Universe would be adequately and consistently driven by the less complex Fundamental Interactions. All observations of Intelligent Behavior would simply be a result of Matter serving as an Intelligent Agent of the four Fundamental Interactions. However, since the four Fundamental Interactions exhibit no Intelligence, there could be no Intelligent Agent. Intelligence itself could not exist.

2) The behavior of all Intelligent Beings would be predetermined. Absent Fundamental Intelligence, the Behavior of all inanimate objects in the Universe is predetermined. This is the very essence of the science behind the Standard Model. All things behave based on a set of Standard Interactions plus Gravity. Every action or reaction of Matter since the beginning of the Universe has been effectuated by the four Fundamental Interactions. No thinking, no decision-making. The Universe has and will continue to sail along happily and predictably based on these Universal laws. If another source of Force is not available, Intelligence would have no fundamental influence over Matter. Therefore, if Intelligent Beings are

only controlled by Fundamental Interactions, all the actions of Intelligent Beings would be preset based on those Interactions.

3) Free Will would be eliminated. A lot of work has been done in the past half century on the Neuroscience of Free Will; and there are several scientific schools of thought with regard to the issue. The concept of Hard Determinism takes the position that Free Will cannot exist, since everything in the Universe is governed by a set of fixed laws, flowing from the Fundamental Interactions. This creates a paradox that cannot simply be dismissed.

Free Will. Absent Free Will, why would Intelligence occur in the first place or develop to such a refined level in higher order animals. One might argue that it was essential for the survival or longevity of the species. However, this is not supported by observation. Alligators, tortoises, and Giant Sequoias all have long life spans and their species have survived for a much longer period than Humans, with arguably more limited Intelligence.

There would be no impetus for the advancement of Intelligence in Intelligent Beings unless such Intelligence provides something fundamentally superior to the other Forces of Fundamental Interactions. Looking at this from another reference point, if the Fundamental Interactions provide for all of the Forces necessary for the development and advancement of Life, then Intelligence never would have entered the picture.

Free Will is the Exercise of Intelligence. We can therefore look at this from an evolutionary concept. If a muscle, limb, or organ is not exercised or utilized to support the requirements or behavior of the living organism, than eventually that muscle, limb, or organ will go away through Natural Selection. Free Will allows us to Exercise Intelligence. If we did not exercise singular decision-making, critical thinking, and creative behaviors, our Intelligence would actually start to deteriorate. Therefore, if you eliminate Free Will, the exercise that drives the development of Intelligence would not exist.

The above quandaries demand that Science at least consider that

Fundamental Intelligence is a Phenomenon that can initiate Forces that are independent of the Four Fundamental Interactions.

INTELLIGENT FORCES

The evidence that Intelligent Beings can initiate Independent Intelligent Forces becomes even more pronounced when we look at the Forces that drive some of the observed behaviors of higher order Intelligent Beings; Humans. By examining several of these Forces we can clearly see additional evidence of Fundamental Intelligence and its role in initiating the Forces that drive Intelligent Behavior. These are not the involuntary or reflex Forces that drive basic life functions. These Forces are initiated by Intelligence, and can be driven by independent decision making of the Intelligent Being.

The Force of Imagination: Imagination is one of the best examples of behavior initiated by Intelligence. Intelligent Beings use the Force of Imagination to drive creativity, the foundation for a broad range of actions; from figuring out new ways to use tools and new tools to use, to producing works of visual, audio, and performing arts. Imagination allows us to find new solutions to problems and to express individual personality. These elements in themselves demonstrate that the Force of Imagination and the resulting creative behavior cannot be predetermined. We can demonstrate this with a common observation.

The music being written by Country Music Star Taylor Swift in 2012 could not possibly have been predetermined in, let's say, 1982. Taylor Swift was not alive in 1982, and neither the matter nor the individual molecular structure that makes up Taylor Swift existed at that time. There was no Taylor Swift DNA, and no way of knowing if Taylor Swift DNA would ever be produced.

Her words and music can be directly attributed to her Force of Imagination initiated by her own Intelligence. Other people, made up of nearly identical molecular matter (the exception being the double DNA molecules) driven by identical Fundamental Interactions did not imagine her songs, or create her words or music. The difference is Taylor Swift's Fundamental Intelligence stimulating her Force of Imagination. We can say the same for the art of Leonardo Da Vinci, the literature of William Shakespeare, the designs of Frank Lloyd Wright, and the inventions of Thomas Edison.

The Force of Imagination is able to use the metaphorical Mind's Eye to conceive or visualize something that has never been thought of before. It is one of the most powerful Forces driving Human Behavior. Attributing the Force of Imagination of Intelligent Beings solely to the four Fundamental Interactions, in the absence of any individually initiated Intelligent Force, is contrary to the observed Behavior resulting from Imagination. The Force of Imagination must flow from another source; Fundamental Intelligence.

The Force of Faith: Faith is another exceptionally powerful Force initiated by Intelligence. The Force of Faith results in phenomenal Behavior, particularly in the highest order of Intelligent Beings. We can demonstrate this with another relatively easy observation with profound consequences on Human Behavior. Specifically, let's look at the Force of Faith and how it plays a major role in the behavior of Humans with regard to Economics.

Faith provides value to Money!

Money is a major influence in the daily lives of individuals, communities, societies, nations, and the world. However, most of the monetary currencies used today have no value in their underlying physical properties; their matter. An exception to this might be gold coins if the value of the coin is based on the value of the gold - the actual market value of the matter. However, most currencies are not backed by an equal value of a fungible commodity. Instead, the value is based on the Faith that the individual or society places in the Authority issuing and backing the currency.

We can clearly observe the phenomenon of Faith as a Force for behavior in Intelligent Beings simply by watching what they do with money. The U.S. Dollar is backed by the full Faith and Credit of the United States. No tangible matter or marketable commodity; just Faith in the United States. However, most of us would willingly accept one hundred dollars in exchange for seventy-five dollars worth of a marketable commodity, such as gold or oil. In such an exchange, the Force of Faith just caused us to exchange something with no "inherent matter value" for something with seventy-five dollars worth of "inherent matter value". This is only a good deal if others share the same Faith in the currency.

Humans willingly accept a wide variety of currencies every day, in exchange for valuable tangible matter, solely based on the Force of Faith. So what happens if our Intelligence triggers a reduction in the Force of Faith in the currency? Immediately, the value of the currency begins to fall. There has been no change in the supply and demand of the tangible products; nor has there been any change in the "value promise" of the currency. It is still backed by the full Faith and Credit of the Authority. The only thing that has changed has been the Force of Faith triggered by Intelligence. Changes in the Force of Faith of the people using Dollars will change the Behavior of the people using Dollars, and result in a tangible change in the value of Dollars. Now that's Force!

We have observed this specific phenomenon numerous times in the past. In a period of less than a decade in the 1980's and 1990's the Soviet Union, a highly advanced nation, experienced a rapid loss of Faith in their currency. Intelligent Beings changed their behavior as a result of the Force of Faith, and the value of the Ruble plummeted. As a result the Soviet economy crumbled, in part leading to the collapse of the entire Nation.

Intelligent Beings can self-initiate the Force of Faith, resulting in changes in Behavior that can impact the entire world. There is no evidence whatsoever that the Force of Faith can flow from the Fundamental Interactions. Such an argument would have to demonstrate that such behaviors were as consistent as the Fundamental Interactions. However, that's not what our observations indicate.

Intelligent Beings can and do initiate the Force of Faith and change their behavior even in defiance of statistical data or a normalized response. In other words, the Force of Faith can act contrary to the Fundamental Interactions of Nature. Therefore, the Force of Faith must be independent of the Fundamental Interactions of Nature. The Force of Faith must flow from Fundamental Intelligence.

The Force of Hope: If the Force of Faith is able to maintain or regulate the values of today, the Force of Hope drives Human Behavior with regard to tomorrow. The Force of Hope is one of the most advanced Forces of Intelligent Beings, and may only be evident in Humans (although this limitation is neither a theory nor hypothesis presented herein). A significant portion of Human Behavior can be directly attributed to the Force of Hope, particular in our observations with regard to how Humans prepare for or demonstrate concern over the Future.

I am not talking about the behavior squirrels exhibit in storing acorns. Those types of behavior are acquired for survival. I am talking about the behaviors that cause us to consider the plight of future generations; that cause us to make sacrifices today for the sake of the long term future of our families and society. We know of no other living thing or inanimate object that demonstrates behaviors that show consideration for the long term (multi-generational) future. Another observed Behavior demonstrates the Force of Hope. Humans record history.

Think about it. Humans have developed not only the capacity but also the desire to record history. No other living thing intentionally records history. Intelligent animals, such as dolphins or chimpanzees, certainly acquire knowledge, retain information in memory, and pass that information on to subsequent generations through direct contact. However, Humans alone have developed the capacity to analyze the events of the past and record the analysis and events for the sake of future study.

There are some who would say that the Universe also records its own history. Fossil records, remnants of energy, and other tell tale signs of the past are indeed "recorded" in the matter of the Universe. However, it would be hard to argue that this is a deliberate action of the thing, living or not.

Others would argue that animals record history by leaving their marks on the land; paths between watering holes; migrational roadway "signs" between grazing locations; or biological markings for a specific range or domain of inhabitance. However, just like with the Universe, leaving a residue of your own behavior does not demonstrate either a capacity or desire to analyze and record history for the long range benefit of the Species. Chimpanzees don't scratch simple messages on cave walls for future generations.

Human development of both the capacity and desire to record history is a result of the Force of Hope. We certainly don't record history for the sake or benefit of the past. There is no reason to record history for the present. The Force of Hope creates the desire to record what we know; from Tradition, Reason, and Experimentation, and pass that along for the good of future generations. This desire has caused us to employ another Intelligent Force, Imagination, to develop the capacity to record history. Desire was driven by the Force of Hope; capability was driven by the Force of Imagination.

The Force of Hope causes us to continue to Learn, to continue to strive to be better, and to make a better future; not necessarily for ourselves, but for the overall Human Condition. It is impossible to deny the fact that the Force of Hope drives Human Behavior. It is also impossible to identify any other similar Behavior in any other living or non-living thing in the Universe.

Therefore, it is implausible to attribute the Force of Hope and resulting observable behavior of Intelligent Beings as coming from the same basic Fundamental Interactions that drive the inanimate matter of the Universe. The Force of Hope must therefore flow from Fundamental Intelligence.

These are just three examples of Forces initiated by Intelligent Beings that drive specifically defined and observable behaviors; behaviors that cannot be attributed to merely the involuntary or reflexive Forces that flow from inanimate Fundamental Interactions and the resulting biological reactions. These forces cannot be attributed to Force Carrier Particles, and therefore fall outside of the Standard Model. The concept of Intelligence as both Fundamental and capable of initiating Forces is clearly demonstrated by the behavior of Intelligent Beings. Thus, the Hypothesis is demonstrated and will be accepted as an Axiom for further discussion.

Axiom 11.3: Forces can be Initiated by the Intelligence of Intelligent Beings.

THE G-PHENOM IS ... PULLING IT ALL TOGETHER

This scientific analysis has taken us a long way, so let's pull it all together. The Big Bang Theory, coupled with the Grand Unified Theory and the Theory of Everything, indicates that a Singular Fundamental Phenomenon created the Universe in a nearly instantaneous explosive event. This Singular Phenomenon was the cause of all other elements of the Universe, and effectuated at least four independent actions as part of triggering the Big Bang. We have defined this Singular Fundamental Phenomenon as the G-Phenom.

Based on the G-Phenom's actions in triggering the Big Bang, we determined that It's behavior is indicative of Intelligence. Since the G-Phenom was a Singularity prior to triggering the Big Bang, this Intelligent Behavior must have been self-derived. It could not have been the result of any interaction or other outside influence. It stands to reason that if the Fundamental Phenomenon that triggered the Big Bang exhibited Behavior that could be defined as Intelligent, then that Intelligence must have been a Characteristic of the Fundamental Phenomenon. Therefore this Intelligence can also be considered Fundamental, supporting the Theory of Fundamental Intelligence.

Finally, we recognized that the G-Phenom initiated a Force. However, again referring to the G-Phenom's Singularity, the Force the G-Phenom initiated could not have been the result of any Interaction, for nothing other than the G-Phenom existed. Since Forces can only be initiated by Interactions or Intelligent Beings, and since the G-Phenom exhibited Intelligent Behavior and initiated a Force, it follows that the G-Phenom is an Intelligent Being.

Therefore we will further refine our Postulate.

Postulate: The G-Phenom is the Singular Fundamental Intelligent Being that Initiated the Creation of the Universe and Triggered the Big Bang.

Chapter 12 - It Just Can't Be!

"Long before the reader has arrived at this part of my work, a crowd of difficulties will have occurred to him. Some of them are so serious that to this day I can hardly reflect on them without being in some degree staggered; but, to the best of my judgment, the number are only apparent, and those that are real are greater not, I think, fatal to the theory."
 - Charles Darwin, *The Origin of Species*, Chapter Six, Sixth Edition, 1876

THE G-PHENOM CAN'T BE!

At this point there will be many who are shaking their head, absolutely refusing to accept the Theory of Supreme Intelligence or the Postulate of the G-Phenom. Some may have given up on this book long before now, tossing it in the trash or fireplace (hopefully not if they were reading an E-Book version). They may find problems with the axioms presented, the alternatives analysis, or the concept in general. Possibly they are so fixed on their position rejecting a Superior Intelligent Being that there is simply no way to accept any analysis that supports it.

In the off chance that someone who refuses to accept either the Theory or the Postulate has made it this far, I offer the following reflections. These reflections may also be of interest to those who concur with the Theory or Postulate, or at least are open minded enough to consider it.

CRITICAL REVIEW

From the beginning of this Engineering Treatise, I have encouraged questioning, debate, and critical thinking on every issue presented. I have attempted to introduce alternate theories and competing hypotheses as part of a rigorous discourse on the Theory that is the basis for this work; and the Postulate derived from the analysis. After all, that's what Science is all about.

Hopefully as you have worked through the various theories, hypotheses, and axioms you have conducted some additional research or discussion as part of your personal evaluation of the Matter of Fact. As discussed in the Chapter on Learning, we all must continue to Learn in order to maintain Balanced Knowledge.

As Humans, we are each provided with the ability to Reason. Whether you believe that you have the Free Will to exercise that Reason or believe that Hard Determinism preordains all Human Behavior, at the very least you should attempt to employ your own Reason in applying Tradition and Experimentation to this study. If your Reasoning causes you to question any of the material presented, then by all means you should research the matter further. Such research could be through individual study or group discussion. Group discussion brings the resources of multiple Intelligent Beings together, expanding the power of Reasoning.

In addition, Tradition, or the thousands of years of previous reasoning and observation should not be thrown out. While current technologies allow our Experimentation Learning to delve deeper into both the micro-World and the macro-Universe, it does not trump all the Knowledge of the past that is provided to us through Tradition. This is particularly evident when it comes to the Interactions of Intelligent Beings and the Forces driving Human Behavior; the understanding of which is still relatively infant.

We can and should base our actions on the Balanced Knowledge we have at the time, recognizing that future Learning will alter that Balanced Knowledge. As we continually Learn, we will find that our Experimentation, our Reason, and our Tradition are all fallible. However, that does not make any of the Knowledge conveyed through those Learning Channels bad; it was the best Knowledge we had at the time.

We must critically review all Knowledge as part of Learning. That's why Learning is Continuous, and our Knowledge must be Balanced.

BAD LEARNING

New Learning will continuously allow us to update Balanced Knowledge. This does not mean the old Learning was bad; it was the best Learning available at the time.

However, there is clearly a phenomenon we can describe as Bad Learning. Bad Learning occurs when absolute restrictions are placed on our Learning such that we cannot employ all three Learning Channels to support Balanced Knowledge. Bad Learning can go so far as to not only restrict Learning to a single Channel, but also restrict the Learning within that Channel to a single school of thought. This phenomenon is often observed in highly restricted Tradition schools, wherein Learning is controlled so that Knowledge is limited to a single dogma or perspective. We have alarming examples of the results of such Bad Learning in the intolerant theocracies that support terroristic Human Behavior.

Bad Learning is not however limited to Tradition. We can also see the results of Bad Learning in both the Reason and Experimentation Channels. This became painfully evident in the rise of the German Third Reich, which had nothing to do with Restricted Tradition. The appeal of Adolf Hitler and his Nazi Party originated from Restricted Reasoning.

Hitler's charismatic style and strong appeal allowed him to convince a large population of highly intelligent people to accept his Reasoning that the Ubermensch concept of Nietzsche was the way to drive society forward. Hitler developed a highly effective propaganda machine to ensure that all the people of Germany "Learned" his reasoning - particularly the youth. Through Restricted Reasoning, Hitler was able to "invent" a master race, demonize another culture, and align an entire country to his restricted, narrow, and dangerous philosophy.

Once Hitler had control over the Reasoning Channel of Learning, he was able to quickly initiate a highly restricted economic

Experiment using centralized planning and a tremendous military buildup. His limited and restrictive Economic Experiment was highly successful in turning the German economy around coming out of the Great Depression (as any big sudden surge in spending would). Just as with the Reasoning Channel, the German people had learned through Experimentation that the Philosophies of the Third Reich had quickly and effectively solved their economic woes. In six short years, Hitler's restricted Reasoning and Experimentation had created one of the most zealous Nationalist Movements the world has ever seen - leading to the most horrific war the world has ever experienced.

Now that's **BAD LEARNING.**

Bad Science. I have tried to take every measure I can to make sure that this book is NOT Bad Science. I have encouraged critical thinking as well as additional research and discussion. I have attempted to present at least some of the alternate hypotheses; yet there are probably a number of others that should also be explored. Finally, I do not expect everyone to accept either the Theory of Supreme Intelligence or the Postulate of the G-Phenom. There can and should be dissenting voices for any scientific theory so that all arguments surrounding the theory can be vetted.

There will be those who want to advance an alternate Theory or Postulate that they believe is more strongly supported by the current body of Balanced Knowledge. For example, some may choose to present and study the theory of an infinite Universe, with no beginning and no dimension; an alternate to the Big Bang. Or there are schools of thought that prefer to look for an unintelligent fundamental triggering event that may emerge from M-Theory. Such alternate theories should be welcome as a matter of scientific study.

At the same time, the Theory of Supreme Intelligence and the Postulate of the G-Phenom should not be eliminated or restricted from scientific study. Both have been presented using a scientific analysis, and while some readers may not agree with the presentation or findings; it certainly does not warrant summarily rejecting the Theory or the Postulate.

The Theory of Everything establishes the possibility of a

Singularity as the explanation of everything in the Universe. This work presents the Postulate of the G-Phenom as a Singular Supreme Intelligent Being that Initiated the Creation of the Universe and Triggered the Big Bang. This Supreme Intelligent Being is also the source of Fundamental Intelligence. This is at least parallel to the concept of the Theory of Everything. Therefore, this Postulate is and should be open for discussion, review, and debate as well as further scientific study.

Excluding the G-Phenom or the Theory of Supreme Intelligence from scientific study is simply Bad Science.

A Blind Eye. A chief culprit in Bad Science is turning a Blind Eye on a subject that a particular scientific school of thought does not find acceptable. The past century or so of scientific study has allowed the deliberate and continuous erosion of the concept of an Intelligent Being as the Creator of the Universe. It has gotten to the point that in some circles even raising such a concept as scientific theory is scoffed at, ridiculed, or considered to be ignorant. Therefore mainstream Science has turned a Blind Eye on this Restriction in Scientific Thought.

Much of this Restricted Science has flowed from the beliefs of confirmed Atheists. These scientists were not neutral on the concept of a Singular Supreme Intelligent Being. They were (and are) specifically opposed to any such concept and provide no allowance for such a phenomenon. Highly respected and studied physicists have significantly restricted the breadth of consideration when it comes to the Phenomenon of Fundamental Intelligence. Unfortunately, we have protected such Restricted Science from alternate theories that would allow for a Supreme Intelligent Being under the guise of Religious Freedom. Ironically, all we have done is restrict Science to Atheism, itself a belief system and religious Tradition.

We cannot continue to turn a Blind Eye to the Restricted Science that prevents the discussion of the G-Phenom or similar theories as possible solutions to the Theory of Everything.

Memorize What I Tell You. This is possibly one of the biggest concerns for the way we teach Science in the United States today; and one of the primary reasons why we must allow discussion of the Theory of Supreme Intelligence and the Postulate of the G-Phenom.

Fifty years ago, Junior High School science class introduced students to the world of exploration, experimentation, and wonderment. To that point in our education, most of what we had learned was through the Tradition Channel - other people's experience. Now, however, we were given the opportunity to open the doors of Experimental Learning. We didn't read the answer in a book; we experienced it. We combined chemicals in beakers, or dissected frogs, or sampled water in streams; in order to Learn through our own experiments and observations. We were introduced to the marvel of Scientific Method.

Our Balanced Knowledge was expanding rapidly, and being a "student of science" meant you were part of it. As a nation we had challenged ourselves to land a man on the moon. Jacques Cousteau brought ocean exploration into the living room. The vacuum tube was yielding to the transistor. Science class was all about learning a new way of Learning; learning by experimentation, observation, and hypothesis.

I can remember learning as much from failed experiments as from successful ones. Being right or wrong was less important than being able to work through the process. You had to learn to define or address a problem, craft a reasonable hypothesis, and provide supportable evidence as to whether the hypothesis was likely true or likely false based on Experimentation, Observation, or Mathematical Logic. Proving a hypothesis wrong can be as important as proving one right.

Unfortunately that has changed. Now we measure students' progress with annual tests that are more about the education system than about a student's learning. We teach Science as a series of defined elements of confirmed Knowledge. There is little wonderment and even less discovery. We are taught what to know, and then tested to ensure we know it. Science is no longer an exploration into the Unknown, but a confirmation of the Known.

The study of the initiation of the Universe and the Creation of life is a prime example of restricting teaching to what we believe is "confirmed knowledge". There are many unknowns when it comes to the questions regarding "Where we came from", and "Why we're here." Of even greater significance is the understanding of the Interactions and Forces that drive Human Behavior; whether we are dealing with individual physical or mental health, or the social challenges of economic or political systems.

Yet, we have restricted the study of Creation to a singular school of thought. We won't let middle school or high school students puzzle over the concept of Fundamental Intelligence. We won't allow students to ponder, as part of scientific wonderment, the G-Phenom, or alternate theories that combine cellular evolution and intelligent DNA interaction. Yet these are clearly viable theories for exploration.

Why Not? Why do we require students to memorize and be tested on a single school of thought with regard to the most fundamental questions of Life? Why is it that Science currently insists that there is no intelligence superior to man's intelligence at work in the Universe?

We must allow scientific study, particularly in the formative years, to explore all possible theories and postulates for the questions and challenges that face us. The last thing we want is to shackle the Experimental Learning of future generations by requiring them to memorize what they are told. Limiting Experimental Learning to only the things we are certain of; or worse, to only the things a few people are certain of; defeats the concept of Experimental Learning. Restricting scientific study to memorizing what we are told is Bad Science.

Is Man the Supreme Intelligent Being? One of the key arguments against the scientific study of a Supreme Intelligent Being and the Phenomenon of Fundamental Intelligence is that there is no evidence that any such Being ever existed. This is one of the clumsiest arguments offered in a scientific analysis; yet for some reason it is effective.

If you want clear evidence of Intelligent Beings, simply look around. Humans, for the most part, meet the definition of Intelligent Beings, and they certainly exist in the Universe. In fact there are several other species in the Animal Kingdom that also demonstrate

Intelligent Behavior. Therefore, there is clear and uncontrovertibly evidence that Intelligent Beings exist.

It is also fairly widely accepted by the scientific community that living things probably exist elsewhere in the Universe, and even acknowledged that such living things may also demonstrate Intelligence. In addition, recent concepts in String Theory indicate that there could be other spatial dimensions outside of those with which we are familiar in the currently observable SpaceTime Continuum. If Science acknowledges that other Intelligent Beings may exist in the Universe; and further that other dimensions may in fact exist; then it must follow that the existence of Intelligent Beings is a possibility outside of the current knowledge of our Universe.

The only question remaining is whether man is the Supreme Intelligent Being in the Universe. We simply cannot make such an assumption. In fact, the existence of Human Intelligent Beings on earth is a relatively recent and isolated phenomenon when taken in context to the entire SpaceTime Continuum. Therefore to assume that man is the Supreme Intelligent Being in the Universe defies scientific reason. It also demonstrates a level of intellectual arrogance.

We cannot assume that man is either the sole or supreme Intelligence in the Universe. There is clearly a chance, if not a likelihood, that other Intelligent Beings exist outside of human observation, possibly within undiscovered dimensions. Therefore, we must allow that a Supreme Intelligent Being could be the Singularity that Created the Universe and Triggered the Big Bang.

This is at least as probable as the existence of an undefined unintelligent singularity that continues to evade our logic or observation. Unless the body of Balanced Knowledge demonstrates that Man is the Supreme Intelligence of the Universe, then the Postulate of the G-Phenom must be allowed as a plausible and valid Scientific Theory, and is further refined as follows:

> **Postulate: The G-Phenom is the Singular Supreme Fundamental Intelligent Being that Initiated the Creation of the Universe and Triggered the Big Bang.**

PART FIVE: Intelligent Life

Chapter 13 - From Big Bang to Birth

THE FUNDAMENTAL NATURE OF INTELLIGENCE

If you accept that the Universe had a Beginning, then it's relatively straight forward to accept that the Beginning of the Universe probably resulted from a Phenomenal occurrence, such as the Big Bang. This Phenomenal occurrence was more than likely triggered by a Singular Fundamental Phenomenon; a concept that correlates with the Theory of Everything.

Starting with the Theory of Fundamental Intelligence, we applied a scientific analysis to develop and propose the Postulate of the G-Phenom, a plausible explanation for the initiation of the Universe and the Triggering of the Big Bang. Along the way, we discussed a variety of Somethings that are Fundamental to the Universe as we know it. Most of us are familiar with the concepts of Matter, Mass and Energy. The discussion on SpaceTime, Interactions, and Force added not only a framework and animation to the static Universe, but also a level of complexity to the model. Much of the theory incorporating these Phenomena can be found in the Standard Model of Particle Physics and Relativity.

However, probably the most controversial concept introduced so far is the Theory of Fundamental Intelligence itself. While substantial observation and reason were presented as evidence supporting this Theory, it will take significant additional study and a vetting of dissenting opinion to advance our understanding of Intelligence.

Innate Intelligence. The concept of Intelligence is certainly not foreign to scientific study. There are volumes of information on the various theories of Intelligence, including what Intelligence is, where it comes from, and how it influences behavior. The study of Cognitive Abilities often deals with difficult questions, such as whether there is one intelligence or many. A specific theory regarding a concept of how Intelligence might drive biological forces is illustrated in the supposition of Innate Intelligence.

Innate Intelligence is not a concept of religious Tradition or abstract Reason. The term emerged from Chiropractic Science. Innate Intelligence is a theory that deals with the internally derived forces of living things that provide intrinsic healing mechanisms. Innate Intelligence recognizes that Intelligence may create forces that are outside of the Forces initiated by the Fundamental Interactions of Matter; Forces that are not attributable to Particle Physics.

Ironically, the concept of Innate Intelligence is considered to be a negative influence on the acceptance of Chiropractic Science by the general medical community - the same community that thought Bloodletting was scientifically sound only a century ago.

Supreme Intelligence. The concept of Fundamental Intelligence, and the existence of a Singular Fundamental Intelligent Being as the Originator of the Universe is an important matter for scientific discussion. It opens an entirely new realm of possibility for the behavior of the Universe.

The concept of Fundamental Intelligence would lead us directly to the notion of Supreme Intelligence; the source of all of the Intelligence in the Universe. We may find that we are able to immediately expand Human Intelligence regarding the Universe by merely accepting the possibility that Humans are not the Supreme Intelligence in the Universe. Once we accept that possibility, the Postulate of the G-Phenom naturally and rationally emerges.

This is another important point. If we allow that something out there in the cosmos or beyond, possibly in a dimension of which we are not even aware, may be more intelligent than we are, it opens up entirely new opportunities for Learning. This can be easily illustrated in basic learning theory. If we believe we are as smart as or smarter than the teacher; it's hard to learn anything. However, as soon as we open our minds to the possibility (or probability) that someone or something might be smarter than we are, we can begin Learning.

The real value in a greater understanding of the Postulate of the G-Phenom and the Theory of Supreme Intelligence lies not only in the understanding of the behavior of the Universe. The real value lies in the opportunity to significantly advance our understanding of the next Phenomenal occurrence in the Universe: **Life.**

AFTER THE BANG

BANG - The G-Phenom has just Triggered a Force that exploded the Mass-Energy Equivalence for the entire Universe, initiating the formation of Matter and accelerating it into the SpaceTime continuum. The Universe has begun.

The Fundamental Interactions now take the lead on providing all the Forces necessary to drive the development of the Universe. As the initial impact of the Big Bang begins to subside, the entire Universe starts to cool down a bit and molecules begin to form, producing Matter and creating elements. The volume of Matter grows, ranging from subatomic particles to planets, stars, and galaxies. This bubbling cauldron of Universe soup was set in motion by a Singular Fundamental Intelligence, and for billions of years the stew continued to simmer under the direction of the Fundamental Interactions.

The G-Phenom, however, did not disappear once the Universe started. It maintained its role as the Singular Fundamental Intelligent Being following the Big Bang. Just as with the four Fundamental Interactions, the G-Phenom remained a potent Phenomenon in the Universe. There is certainly no reason to believe that a Singular Fundamental Intelligence, upon Triggering the most phenomenal thing in the Universe - the Universe itself; would then simply disappear, or somehow break itself down into Nothing. Further, it doesn't stand the test of reason to assume that Fundamental Intelligence would replace itself with four non-Intelligent Fundamental Interactions.

Steered by the forces of the Fundamental Interactions, the Universe continued to expand and matter started to find a point of Equilibrium. The Fundamental Interactions drove portions of the Universe into a fairly balanced state, and relatively stable solar systems were created. About 4.5 billion years ago, in the Milky Way Galaxy, a medium sized planet took shape, occupying a consistent orbit around a large star. The Earth emerged.

At this point, we have not considered any additional action or Behavior by the G-Phenom following the initial Triggering of the Big Bang. Current Scientific theories attribute all forces driving the Universe forward after the Big Bang solely to the four Fundamental Interactions of Matter. Restricted Science currently makes no allowance for any other Fundamental Interactions, Forces, or Phenomenon, intelligent or otherwise. Based on this limitation, we are forced to consider that the Fundamental Interaction of Gravity provides all of the force for the stabilization of all matter in the macro Universe.

This Restricted Science leaves us with two really big dilemmas; The Dark Universe and the Expanding Universe.

The Dark Universe. Which of the following best describes the scientific theory of the Dark Universe?

1. It's the world of Lord Voldemort
2. It's the playground of Darth Vader
3. It's the haberdashery for Dark Helmet
4. It's the inferred but never observed or detected matter and energy that make up 95% of the Universe

Our current understanding of the Fundamental Interaction of Gravity does not account for some of our celestial observations; particularly relative to the velocities of orbiting planets in outlying galaxies. Instead of allowing that our understanding of Gravity may be in error, or that there may in fact be a fifth Fundamental Phenomenon at work, current scientific theory has "plugged" the gap between our observation and our knowledge with the Dark Universe - a combination of Dark Matter and Dark Energy.

As discussed previously, the Dark Universe is not a small plug, or an adjustment for rounding error. If the hypothesis of the Dark Universe is correct, then Dark Matter and Dark Energy make up about 95% of the entire Universe. Yet we have no direct evidence that any of this 95% of the Universe even exists.

The only reason for the Postulate of the Dark Universe is that without it, science must consider a new theory or hypothesis that allows for another Fundamental Phenomenon; a Phenomenon such as the G-Phenom. However, current Restricted Science is so dead set against such a theory that it would rather assume that 95% of the Universe is simply undefined, undetected, and unobserved. If, however, the Fundamental Phenomenon that Triggered the Big Bang continued to exist after the Big Bang; and it likely did; it could certainly offer a reasonable alternate to the Dark Universe.

The Expanding Universe. The concept of an Expanding Universe is fairly well accepted by a large portion of the scientific community. This theory is based primarily on Hubble's Law and the concept of redshift in light. Redshift is the definition for what we observe when light moves away from us at a very high speed. The light moving away causes the wavelength to change; the Doppler effect. This is the same effect we experience listening to a siren or horn on a fast moving vehicle; and the same technology used in radar based speed detection devices. If we observe redshift, then the light is moving away from us; blue shift, the light is moving toward us.

Current earth observations have detected redshift in the light received from distant galaxies. This is indicative of a Universe that is expanding in all directions from the point of observation. The question is whether this Expansion is constant, decelerating, or accelerating. It also raises another important consideration; where is the center of the Expansion (this is a subject for another study)?

Our initial reaction might be that once Matter was flung into the SpaceTime Continuum by the Big Bang, there was and is nothing to slow it down. Gravity caused matter to coalesce into galaxies as it flew through space at exceptionally high speeds. These massive constructs of matter, behaving as a single mass, have tremendous inertia that cannot be overcome by the Fundamental Interactions.

Therefore, absent any other Force, they must continue in motion at a steady velocity, and therefore Expansion would be constant.

Another reasonable hypothesis might consider that large Interior Galaxies, those closest to the center of Expansion, create a big enough deflection in the SpaceTime Continuum to provide an attractive Force to the outlying Galaxies. Large galaxies such as the Milky Way and Andromeda might actually be slowing the Expansion down and pulling things back together. This hypothesis is further supported by the fact that our observations of Andromeda indicate a blue shift, and therefore Andromeda may actually be coming closer to Milky Way (Earth's home Galaxy). Based on this hypothesis, at some point the velocity of outlying galaxies would drop to zero, and then start heading in the opposite direction. This would mean that the Universe's Expansion is decelerating, and at some point it will flip to contraction.

Here's the problem with both of the above theories. Our latest observations and calculations suggest that the Universe's Expansion is actually accelerating. However, this is inconsistent with both Gravity and the SpaceTime Continuum. Any galaxies at the outlying edge of SpaceTime cannot be influenced by any Mass beyond their position - they are at the edge. There would be nothing beyond these galaxies that could bend SpaceTime. Yet something must be overcoming the inertia of these galaxies' mass, since they appear to be accelerating.

By definition, you cannot accelerate a mass without a force. If the Universe's Expansion is indeed accelerating, then there must be another Force overcoming the inertia of the galaxies. Some claim that Dark Energy is the force for accelerating galaxies. But energy is not force. Something has to convert energy to force. So, we are left with another unanswered question; Unless …

THE G-PHENOM EXISTS!

There is little question that a Force Triggered the Big Bang. By definition the initial Mass-Energy Equivalence could not have changed its behavior and expanded into SpaceTime without a Force.

Additionally, it seems reasonable that whatever initiated the Force that started the Universe could exist after the Big Bang. There's little reason to think otherwise. If this initial Force was powerful enough to start the Universe, it would certainly be powerful enough to continue to accelerate the Expansion. In addition, if the Force had been initiated by a Supreme Intelligent Being, the G-Phenom; then it could certainly cause the behaviors we have observed, such as gravitational lensing.

Limiting scientific study to the four Fundamental Interactions of Matter as the source for all Forces in the Universe clearly creates a conundrum with regard to it's behavior during the first ten billions years of expansion. However, the biggest challenge to Restricted Science is about to step onto the stage. The first indication of a new, previously undetected and totally unprecedented Phenomenon is about to appear on that same small planet in the Milky Way that was now at its billionth birthday – Earth.

THIS NEW PHENOMENON IS LIFE.

Chapter 14 - The Birth of Life

LIFE BEGINS

The first indications of Life on Earth appear in fossils dating back about 3.5 billion years. This of course does not mean that Life could not, did not, or does not exist elsewhere in the vast Universe. In fact, many in the scientific community feel that Life could very well exist elsewhere, and may have formed earlier than Life on Earth. While we should certainly allow for that possibility, at this point we have not observed evidence of Life in any other form in any other location. Therefore, all of our observations with regard to Life must be based on Life on Earth.

The scientific analysis of the Creation of the Universe is indeed challenging. Scientific study of the Creation of Life increases the complexity by an order of magnitude. Identifying the Forces and Interactions instrumental in the Origin of Life is at least as bewildering as the initiation of the Universe. We have an even bigger pool of variables, a larger number of unknowns, and a more complicated set of permutations in the Creation and Development of Life.

A review of what we know about the Science of Life may shed additional light on both the Theory of Fundamental Intelligence and the Postulate of the G-Phenom.

Warning: If you think the behavior analysis of the Big Bang was tedious - stand by. This discussion will be simply mind-boggling. However, since this is an Engineering Treatise I will again take a simple approach to analysis (simple mind boggling); partly to keep this Book from evolving into a thousand pages, and partly because I simply don't know any better.

Caution: I must reiterate that this Book is not intended to provide answers. Too many books provide all the answers, particularly with regard to Life. This book is intended to spur additional question, discussion, critical thinking, and of course, Scientific Study. We must keep Learning so that we can maintain Balanced Knowledge on the subject of Life, and the critical Forces that drive Life's Behaviors.

WE KNOW AND WE DON'T KNOW

We don't know the Origin of Life. We don't know what caused it to begin, or the initial process by which cells started to form. We do know that all life is made up of cellular structures, but we don't know why bacteria doesn't have a nucleus; or why plant cell nuclei and animal cell nuclei are different.

We know that Life must have a number of ingredients to exist. Every form of life on earth, with the exception of viruses (which are not independent of other life), requires three Macromolecules; Protein, DNA, and RNA. These macromolecules are complex in their own right, and we don't have a good answer as to how any of them came into existence alone; much less in some combined and interactive state. Scientists have hypothesized that RNA came first and may have initiated the others; yet we have been unable at this point to confirm that RNA could have been the fundamental element to Life.

Even if RNA is the fundamental initiator of Protein and DNA, we are left with complex challenges in things like cell membranes and genes. Genes depend on Proteins, and Proteins depend on Genes. So both have to come on the scene at the same time. Of course for Life to exist, all of these things would have had to occur and be packaged in a nice, organized Cell, which contained the ability to generate the Forces and process to split itself in two.

This is another big point (so write this down)! Life has the internal Force necessary to drive a self sustaining Behavior - Cells divide. We do know that Fundamental Interactions can allow atomic structures to break down, or transform from one state to another. However, other than living cells, we know of no other thing in the Universe that has the ability to simply Create a duplicate of itself!

Vitalism. I want to draw a distinction between the Theory of Fundamental Intelligence and Axiom of Intelligent Forces, and the life theory of Vitalism.

Vitalism is a concept of Life that at first may be confused with or compared with The Theory of Fundamental Intelligence. However, it is based on entirely different principles. Vitalism fell into disfavor in the Scientific Community over the last two hundred years as biological functions of life were demonstrated to be attributed to the Fundamental Interactions of Matter. However, the evidence leading to the rejection of Vitalism actually strengthens the argument for the Theory of Fundamental Intelligence and the Axiom of Intelligent Forces.

As stated previously, it is apparent that many of the Forces that drive the biological behaviors of all living things are derived directly from the Fundamental Interactions of Matter. What is equally clear is that some of the observed behaviors of Intelligent Beings cannot be attributed solely to those same Fundamental Interactions. There is clear evidence that Intelligent Beings use Intelligent Forces to also drive behavior.

The Fundamental Theory of Intelligence does not replace the Fundamental Interactions of Matter in driving the Behavior of Intelligent Beings: It augments them. Fundamental Intelligence is a separate and distinct Fundamental Phenomenon that joins with the four Fundamental Interactions of Matter to establish the Natural Forces for all Intelligent Beings. The Matter of the Physical Form of all Living things is driven by Fundamental Interactions. Some of the observed Behaviors of Intelligent Beings, however, are driven by a fifth Fundamental Phenomenon; Fundamental Intelligence.

I told you your head would hurt. With all that we don't know, what do we know?

We know that Life in fact exists. We know that something triggered the Origin of Life on Earth. We have proof of that by the fact that Life Exists.

However, for some reason Restricted Science refuses to even consider that Life on Earth may have been a result of a Force outside of the Fundamental Interactions of Matter. As with the study of the beginning of the Universe, Science currently limits the study of the Origin of Life to another random, spontaneous event that flows solely from the Laws pertaining to Matter.

If a Force other than the Fundamental Interactions initiated the Universe, certainly such a Force could have Originated Life. The Theory of Supreme Intelligence provides for such a Force. The Postulate of the G-Phenom provides a reasonable answer to a significant number of the open questions. Scientific study of these concepts offers plausible solutions to some of our most fundamental questions regarding Life. Science must at least consider them.

LIFE TICKS

To many of us, Life is the most important thing in life. However, within the Universe, Life is a very small piece of a very large puzzle; both in terms of physical dimension as well as the expanse of time. The Cosmos revolves around the interplanetary and galactic forces that drive the Universe, with little regard for Life. The Universe could, and has, existed without Life - but Life would not exist without a Universe.

Despite this fact, our entire perspective of our Universe is based on the singular reference from which we all observe everything - Life. But what is Life? What is the common bond we as humans share with an oak tree, or protozoa? More importantly, what makes Human Life unique from all other life forms that inhabit our world?

Life, both its Origin and its Evolution, has been a perplexity that man has pondered since the beginning of recorded history. We have always wondered where we came from and where we are going. The first step to answering this problem is figuring out what life is, and what makes life work.

Starting with the basics, all life is made up of matter and energy. However, inanimate or "lifeless" objects are also made up of matter and energy. A lump of coal is an organic piece of matter that certainly contains energy. That piece of coal may be the remnants of a past living thing; however, it is most certainly not alive. So, we need to look further to find the discriminator between "alive" and not "alive".

Scientific study has allowed us to acquire an amazing amount of knowledge with regard to how living things work. There are volumes of information constantly being augmented by new discovery on organisms, DNA, cell structures, metabolism, and the flow of energy within living organisms; down to the transfer of neurons between synapses. The Human Genome offered another giant leap forward in the understanding of human life. But just as with all scientific endeavor, the more we know, the more questions we are left with.

There is clear indication that there is more to life than simply Matter and Energy. This is particularly true for Intelligent Life, since as discussed previously the behaviors of Intelligent Beings cannot be completely attributed to the Fundamental Interactions of Matter. After all, if life were simply an electrical charge passing through an appropriate set of ingredients, it could spontaneously erupt from an accidental mix of those inanimate ingredients. Yet we have no evidence that Life has in the past or is currently "erupting" from any such accidental mixing of ingredients or the theoretical Primordial Soup.

The earth is in one of its most prolific stages of supporting and regenerating Life. Yet, in all of Human study and observation there is no evidence that any new living cells, organisms, or other entities have sprung up from or emerged from any non-living ingredients. There isn't even any evidence of any other thing in the Universe, other than a living cell, that has the ability to Create an exact duplicate of itself – no other atom, molecule, or subatomic particle. Only the living cell can perform this activity.

One of the scientific arguments for the lack of evidence of new erupting life is that such an eruption of life from non-living ingredients would only have occurred at a given period of time, about 3.5 billion years ago. As the hypothesis goes, conditions for the accidental and spontaneous eruption of Life were just right at that time.

However, there is no evidence to support such a hypothesis or to demonstrate that any such set of perfect conditions in fact existed. Indeed, the primary argument in support of such a hypothesis is that the conditions must have been perfect since that's when it appears that life in fact first erupted. It's a circular argument that must be made to support the totally accidental and random eruption of life; and to dismiss any other Fundamental Phenomenon that may have played a role in the initial Creation of Life.

Understanding Life is more than acquiring knowledge regarding the matter that makes up the Physical Form of Living things, or the interactions that support the basic biological functions. Life is not simply a fabric of chemical reactions. Our attempt to study life as the sum of numerous component parts or divisible functions also leaves huge gaps in our understanding of Life Systems and Life Behaviors.

Life is dynamic, and only exists in the dimension of time. Without time, there is no Life. In order to understand Life and the things that cause Human Behavior and Interaction, we need to evaluate the Forces that drive life forward in time; in other words, we need to know what makes Life Tick.

LIFE: ORIGIN AND DEVELOPMENT

One of the most important subjects for scientific study is the Behavior of Living things; and in particular Human Behavior and Human Interaction. Such study leads to advances in everything from medicine, to economics, to politics. The best way to analyze the Forces that drive Human Behavior is to delve into the mysteries surrounding the Origin and Development of Life, leading of course to Human Life.

Therefore, get ready for a really rough ride as we approach the scientific study of the Origin and Development of Life through two very familiar but often explosive terms; **Evolution** and **Creation**.

Chapter 15 - Evolution and Creation

A MATTER OF TERMS

Evolution and Creation. Why is there so much consternation over two common terms, particularly in the context of Life? To some the concept of Evolution is completely contrary to their faith Tradition, threatening their entire belief system. To others, the thought of Creation is simplistic and naive, and has no basis in scientific study. As with many other areas of study, this problem with terminology arises when we restrict our understanding or critical thinking on issues to a single view or reference point.

For centuries, Restricted Tradition left no room for the new discoveries of Science. The studies and findings of Copernicus, Galileo, and Darwin, to name a few, were soundly rejected by Restricted Tradition, and scientists were often persecuted for offering theories contrary to the established belief system. In the last two hundred years, the tables have completely turned. Now, it is Restricted Science that rejects and ridicules Tradition Learning with regard to the Origin of the Universe and the Creation of Life.

What if, however, these terms and the theories they represent are NOT mutually exclusive. Can Evolution and Creation concepts be complimentary?

The only way to find out is to scientifically study the fundamental meanings of the two terms, without allowing Linguistic Determinism to limit our thinking. If instead of focusing on the differences in these theories, we open our minds to their intersection, we might actually advance our Balanced Knowledge; we might Learn something?

We won't know unless we try.

EVOLUTION

The Cause of Evolution

Charles Darwin's Work. There is little doubt that evolution occurs. Thinking regarding the evolution in plants and animals dates back many centuries. Charles Darwin's work, and particularly his book *The Origin of Species* anchored the Theory of Evolution as a fundamental principal for the development of Life. Since that time, the advances in Evolutionary Theory have been phenomenal. Recent decades of observation and experimentation have allowed Science to unlock the secrets of heredity; genes, chromosomes, DNA, and the Human Genome.

However, as rock solid as Evolutionary Theory appears to be, there are still many gaps in our observations and understanding that leave large questions. In addition, for some reason we have limited the study of the Evolution to a single process, Natural Selection. This is contrary to both Charles Darwin's observation and stated conviction.

> "...I am convinced that Natural Selection has been the
> main but not exclusive means of modification."
> -Charles Darwin, *The Origin of Species*

Indeed Darwin begins his study on Evolution through the analysis of another very powerful form of Selection; a form of Selection that is almost completely ignored in our current teaching of Evolution. Could this be a case of Restricted Science?

We will address this other powerful form of Selection in a few moments. As to Natural Selection, Darwin himself raised several challenges to the issues in *The Origin of Species.* In Chapters 6 and 7 of that book, he faced head on a number of the difficulties with the Theory and put some of them to rest. Other more obtuse questions remain open to this date.

For example, a major paradox in the Theory of Natural Selection is the conundrum of Altruism in animals demonstrating a high level of intelligence. Altruism is significantly different than and should not be confused with Cooperation. Natural Selection may clearly play an important role in the genetic disposition of some species to cooperate. Cooperation is a wide-ranging behavior observed across many species, from ant colonies to chimpanzee families.

Altruism is a completely separate behavior. By definition, altruistic behavior causes someone to sacrifice something of value with no expectation of any return benefit. Again, this is not the cooperative behavior we see within a social group; it is behavior that reaches far outside the group. Altruism is completely contrary to the concept of survival of the fittest.

Another major issue for Darwin was the Evolutionary impact of a fertile cross species Hybrid, particularly in non-domesticated animals. If such fertile hybrids existed in significant numbers, Darwin realized it would have a tremendous skewing effect on the data used to track the Natural Selection of Living Things. In Chapter Eight of *The Origin of Species* Darwin deals with this concern in detail. The following passage indicates the concern.

> "If our systematic arrangements can be trusted, that is
> if the genera of animals are as distinct from each other,
> as are the genera of plants, then we may infer that
> animals more widely separated in the scale of nature
> can be more easily crossed than in the case of plants;
> but the hybrids themselves are, I think, more sterile. I
> doubt whether any case of a perfectly fertile hybrid
> animal can be considered as thoroughly well
> authenticated."
> -Charles Darwin, *The Origin of Species*

However, we now know that fertile cross species hybrids do in fact exist. There are numerous Bovid Species, derived from hybrids, which are now highly successful. However, in order to illustrate this issue, we will start with a much simpler example of a hybrid: Mules.

Why Mules? Because Mules are an example of an animal that DID NOT EVOLVE by Natural Selection, and are considered in Darwin's observations. This may be a surprise to all those who are committed to and adamant that every living form or variant evolved from a previous living form by Natural Selection. This is simply not true. So why do we teach that concept in Science and Biology today? Let's repeat that statement.

Mules DID NOT EVOLVE or develop by Natural Selection. In addition, Molly Mules, or female Mules with the capability to reproduce, though rare, do exist and have reproduced. You will see why this fact is important with regard to the Evolutionary process shortly.

For now, let's get back to the earth shattering fact that Mules did not evolve by Natural Selection. In fact, there are a number of Species that exist today, both in the plant and animal kingdoms that did not evolve in the very slow, multi-generational, Natural Selection process that most of us consider Evolution. This is not a parlor trick or question of semantics. The fact is that Darwin recognized that there are two distinctly different Evolutionary Processes that cause the introduction of new species as well as variations within a species. You don't have to believe me; simply read Darwin's work.

For those who don't want to spend time doing that, I will try to provide a quick introduction into Mr. Darwin's study that will clearly illuminate the process of development of Species and variants. In *The Origin of Species*, Darwin contrasts and compares the two primary causes for the development of species and variations in plant and animals; **Natural Selection** and **Man's Selection**.

Natural Selection

The cause of species and variations in plants and animals that we almost singularly refer to and teach as Evolution is Natural Selection. Natural Selection is a process by which the biological characteristics of a plant or animal can, over many generations, slowly change to allow for the most successful characteristics to thrive, while unsuccessful characteristics slowly die off. The initial Change Agent in Natural Selection is a Random Mutation at the cellular level. However, whether that mutation survives through generations of offspring is dependent upon the Natural Selection process. Natural Selection allows the most valuable random mutations; those mutations that lead to the highest success for a Species to compete, survive, and even thrive; to become dominant in the population within the community.

When the Theory of Evolution is discussed, studied, or critically analyzed, it is most often if not always in the context of Natural Selection. That is a major focus of *The Origin of Species*. Unfortunately, the excitement of this new stream of scientific discovery coupled with the tendency toward Restricted Science, has left the impression that Evolution through Natural Selection explains all of Life, including all of the varied species, biological structures, and physiological characteristics.

This is not the case. In fact, Darwin himself recognized another powerful cause of variation in both plants and animals. In the Introduction to *The Origin of Species*, Mr. Darwin states:

> "From these considerations, I shall devote the first chapter of this Abstract to Variation under Domestication. We shall thus see that a large amount of hereditary modification is at least possible; and what is equally or more important, we shall see how great is the power of man in accumulating by his Selection successive slight variations."
> -Charles Darwin, *The Origin of Species*

Darwin's above statement is obvious; yet profoundly changes the game with regard to the entire study of Evolution. Look at how Darwin introduces the thought process that led to his theories; what we so often ignore in our study of Evolution.

Charles Darwin observed an Intelligent Agent (man) utilizing a Selection Process in the Creation of Species or variants; a Process that has nothing whatsoever to do with Natural Selection.

In case you missed it, I shall emphasize Darwin's point in the Introduction to his work:

> **"... how great is the power of man in accumulating by his Selection.."**
> -Charles Darwin, *The Origin of Species*

In Chapter 4 Darwin refers to man's Selection as Artificial Selection; what we now consider as inclusive of Selective Breeding. In fact, Darwin devotes a significant part of The Origin of Species to variation under Domestication, and the Selection process associated with Domestication.

Man's Selection (Intelligent Selection)

What Darwin referred to as Man's Selection, or Artificial Selection, and what now includes the process of Selective Breeding, we will incorporate as a single defined term; **Intelligent Selection**. The justification for combining these descriptors of Selection under the single term "Intelligent" is because they are triggered by a decision of an Intelligent Agent. Under Intelligent Selection, an Intelligent Being makes a decision, no matter what the level of Intelligence or motive of the Being, and triggers a specific process that results in the instantaneous Creation of a variation within a species, or a new species altogether.

Intelligent Selection is just the opposite from Natural Selection. Remember, Natural Selection is triggered by random mutations that must survive through multiple generations to effect a population of a community. Intelligent Selection can introduce major changes in a Species, or an altogether new Species, in a single generation. Further, Intelligent Selection is not artificial: There is nothing artificial about it. Intelligent Selection is a naturally occurring activity based on Intelligent Forces versus the Forces of Fundamental Interactions that drive Natural Selection. With regard to Intelligent Selection, Darwin stated:

> "The great power of this principle of selection is not hypothetical,"
> -Charles Darwin, *The Origin of Species*

Darwin recognized and evaluated in great detail the ability to produce highly specialized species very quickly through Domestication. He clearly delineated the line between species that evolve through Natural Selection, and species that are produced suddenly through Intelligent Selection. Darwin further understood that through the power of intelligence, man could create a wide variety of species and variants that where highly useful and would fulfill a specific purpose nearly perfectly.

> "One of the most remarkable features in our
> domesticated races is that we see in them adaption,
> not indeed to the animal's or plant's own good, but to
> man's use or fancy."

Further;

> "We cannot suppose that all the breeds were suddenly
> produced as perfect and as useful as we now see
> them; indeed, in several cases, we know that this has
> not been their history. The key is man's power of
> accumulative selection: nature gives successive
> variations; man adds them up in certain directions
> useful to him. In this sense he may be said to make for
> himself useful breeds."
> -Charles Darwin, *The Origin of Species*

Darwin demonstrated that there are two fundamentally different Selection processes driving Evolution. The process of Natural Selection is random and is for the survivability or proliferation of the species of variant. The other equally powerful and Natural process is that of Intelligent Selection, whereby species or variants are produced not solely for the benefit of the species, but instead to please or benefit an intelligent being.

Intelligent Selection introduces another profound phenomenon into the entire question of Life. While this phenomenon is exceptionally controversial and therefore volatile, it simply cannot be ignored. This highly contentious issue must be introduced into the scientific discussion of Evolution, since Mr. Darwin introduced the phenomenon as the basis for his Theory of Evolution through Natural Selection. So, are you ready for some fireworks?

Darwin's understanding of the power of Man's Selection in the development of plants and animals; what we are now defining as Intelligent Selection, is a clear and unequivocal demonstration of the

Phenomenon of Intelligent Design!

Intelligent Design

The entire process of Domestication, or Man's Selection, is based on the Intelligent Design of animals and plants. We have already introduced one such product of Intelligent Design, the Mule. There is no doubt that the Mule is a crossing of two Species that was created by Intelligent Selection. The Mule has specific characteristics desirable to support Human requirements. Hence, the Mule may be designed and produced for Man's benefit through Intelligent Selection.

We can see Intelligent Design much more clearly in the Intelligent Selection process used in a much larger Family of Animals, Bovidae. The Bovid Family includes nearly 140 species of ruminant, cloven-hoofed animals. This very diverse Family of animals is found throughout the world, and includes everything from buffalo, antelopes, and gazelles to goats and cattle. While a number of the Bovid Species may have resulted from Natural Selection, there is no doubt that a large portion of this diverse Family were a result of Intelligent Selection driven at least in part, by Intelligent Design. Species and variants of Species, such as sheep, goats, and cattle have all been the subject of Intelligent Selection as animal types were Designed for a specific purpose, or to exhibit specific characteristics.

Another very important observation in the success of Intelligent Selection in the Bovid Species is that while initial hybrids may have been primarily infertile, there were clearly enough fertile specimens for the Species to eventually reproduce as a single Species, and not as a cross Species Hybrid. Domestic cattle are a prime example of a Bovid Species that were introduced, improved, and thrive, even on the open range, as a result of Intelligent Design.

Intelligent Design cannot be denied. Humans acting through Intelligence have designed both plants and animals that would not exist without such Intelligent Selection. The open issue is the variety of Intelligent Agents that may trigger Intelligent Selection. Are Humans the only Intelligent Agents that employ Intelligent Selection?

The Theory of Evolution certainly includes the process of Natural Selection. However, if we are to truly and factually analyze the development of Life, we must also study the other Evolutionary process introduced by Darwin, that of Intelligent Selection. Such study must also include the process by which many Intelligent Selection choices are made, Intelligent Design.

So, let's shore up the argument regarding Intelligent Selection and see if we can draw some conclusions from this discussion.

Darwin's Theory. We have already seen that Mr. Darwin recognized the validity and power of what we have defined as Intelligent Selection, under various terms including Man's Selection, Domestication, and Artificial Selection. In fact, Darwin used his observations of Intelligent Selection to support his Theory of Natural Selection.

> "Can the principle of selection, which we have seen is so potent in the hands of man, apply under nature? I think we shall see that it can act most efficiently. Let the endless number of slight variations and individual differences occurring in our domestic productions, and, in a lesser degree, in those under nature, be borne in mind; as well as the strength of hereditary tendency."
>
> - Charles Darwin, *The Origin of Species*, Chapter 4

Darwin also believed that Intelligent Selection created more variation and diversity within a species than Natural Selection.

> "When we compare the individuals of the same variety or sub-variety of our older cultivated plants and animals, one of the first points which strikes us, is, that they generally differ much more from each other than do the individuals of any one species or variety in a state of nature."
>
> - Charles Darwin, *The Origin of Species*, Chapter 1

It is clear that Darwin considered Intelligent Selection an exceptionally powerful Force in the development of Species and the variations that may occur within a Species. There is also an important nuance in Darwin's various statements comparing Natural Selection and Man's Selection. He allows that Natural Selection is evident from the Intelligent Selection triggered by Man. Darwin opens another profound area for consideration; that Nature can do what Man does. From Darwin's observation we can infer that Nature can also drive Intelligent Selection!

> "Why, if man can by patience select variations most useful to himself, should nature fail in selecting variations useful, under changing conditions of life, to her living products? What limit can be put to this power, acting during long ages and rigidly scrutinising the whole constitution, structure, and habits of each creature, -- favouring the good and rejecting the bad?"
> -Charles Darwin, *The Origin of Species*

What isn't clear is why in the study of Darwin's Theory of Evolution his observations on Man's Selection or Artificial Selection are not presented along with Natural Selection. Further, eliminating the study of Intelligent Selection as an Evolutionary process from the scientific teaching of Evolution is a patent example of Restricted Science.

Molly the Mule: We shall now revisit the case of Molly the Mule. Remember the statement that Molly the Mule was not a result of Evolution through Natural Selection. A Mule is the cross between a Male Donkey and a Female Horse. Note that there is gender specificity in creating a Mule. There is no Natural Selection in creating a Mule; no random mutations that had to be sorted out over thousands of years of genetic strengthening. Instead, a really feisty Donkey meets an appealing Mare and suddenly, magically, a Mule is Created.

The Mule has a number of desirable characteristics that allow it to survive and perform work. However, as a species the Mule's ability to thrive is limited by one simple fact; most Mules cannot reproduce. In fact, there is no recorded instance where a male Mule has ever produced offspring.

The same cannot be said for Molly. Female Mules on rare occasions possess the capability to reproduce. So, assuming either a Male Horse or a Male Donkey finds Molly interesting, another offspring may occur that jump-starts our new hybrid Species in the development line, bypassing Natural Selection completely. In this case, Natural Selection did not Create the Species, but steps in only after Creation to strengthen it and help ensure its survival, or allow its demise.

Let's see how this type of Species Creation might work in a hypothetical Equidae community grazing around in a particular region. In a large population of Donkeys and Horses cohabiting a common range, there would be a high probability that at some point the two species would cross, and Mules would be produced. The most viable reason for such crossing of Species would be an imbalance of male and female populations in one of the two Species, which would attract the attention of the other. There of course could be other reasons, but for purposes of this discussion, a single valid reason will suffice.

Now, since we are dealing with a large population of both animals, we can assume that some number of female Mules would be fertile ("Molly Mules"), albeit a very small percentage. This is what we observe in domestic reproduction. With one or more Molly Mules grazing within the mixed population, there is a high probability that either the male horses or male donkeys would impregnate any Molly Mules. Therefore, within only one or two generations a new Species would be introduced. In addition, as this new Species distances itself genetically from its hybrid Mule heritage, it could become more prolific in reproduction, leading to expanded Species populations.

In the above scenario, neither Molly the Mule nor Molly's offspring, nor the entire population of the new Species is a result of Natural Selection. This entirely new Species is a matter of Intelligent Selection. But whose Intelligence was at work? Good question. Possibly the Male Donkey made an intelligent decision to reproduce with a Horse since no female Donkeys were available. Or possibly another Intelligent Agent is at work. We simply don't know at this point; and therefore this question should be an issue for further scientific study.

The above example is of course, hypothetical. I am neither theorizing nor supposing that a new species of Mule has been created on the open range. However, when you look at the wide variety of species in the Family Equidae, it is highly probable that both Intelligent Selection and Natural Selection have had a role in the development of this Family, and the variety of Species it now includes.

We could suppose that in the early years of any Order, Family, or Genus, the limited number of Species within that group probably increases the likelihood of Intelligent Selection that crosses Species or even Families. Once a new Species is introduced, it will evolve through Natural Selection based on the environment, the habitat, or the long term behaviors of the group. As these developing Species become more specialized and separated; there is probably less tendency for Intelligent Selection to create additional Species; as the forces of Intelligent Selection focus on refinement of Species through slight variations. The process of Natural Selection will determine which of the Species or Variants will ultimately be successful; and could also introduce random mutations that may lead to a new Species.

This supposition of the Intelligent Selection process in the development of a variety of survivable Species within a Family is not just a theory: It is observed behavior. We have in fact already introduced just such a Family; the Bovids.

GMO's. In the last twenty years we have seen an explosion in new variants through Intelligent Selection using Intelligent Design. However, in our desire to align with Restricted Scientific thinking, we have developed a new term for this process and the variants it produces. They are called Genetically Modified Organisms, or "GMO's". The Intelligent Agent in this process is Man. Ironically, the refusal to refer to GMO's as a product of Intelligent Selection or

Intelligent Design leads to a major marketing problem that shouldn't even be an issue.

Ever since Humans began domestication of plants and animals, they have used "genetic engineering" in the Intelligent Selection Process. We have taught such genetic engineering in 4H Clubs for decades. We just called it another name. No matter what name you use, however, the process of Intelligent Selection based on Man's Intelligence and Design has been around for hundreds of years. Advancements in technology have merely allowed Human Intelligence to be more discrete and even more "selective" in the Process of Intelligent Selection. This is exactly what Mr. Darwin observed and predicted.

Nearly everything we eat today that flows from any domesticated production is the result of Intelligent Selection – genetic modification. We have no problem consuming genetically modified variants achieved by grafting plant limbs. We've produced and eaten a wide variety of hybrid apples, for example. However, as soon as we started calling the products Genetically Modified Organisms, a loud outcry is heard.

So let's ponder this question? The pesticides and herbicides we spray on our plants or the antibiotics we inject in animals are the result of relatively recent scientific discoveries in chemistry, biology, and botany that rely on the Fundamental Interactions of Matter. They artificially interrupt natural processes in order to achieve a desired production result, such as killing weeds or reducing infectious diseases in feedlots. Contrast this with the Intelligent Design modifications we make to plants and animals using Intelligent Selection. This is a very Natural Phenomenon that Humans have employed for many centuries to improve the hardiness, quality, and quantity of the food supply. Intelligent Selection has been and continues to be a valuable tool for the development of Life.

So, why is it that we have come to accept the "safety" of products produced through recent scientific discovery that comes through the Fundamental Interactions of Matter (pesticides and herbicides), but we view with great suspicion new products Intelligently Designed and produced through Intelligent Selection (GMO's)?

Could it be our reluctance to embrace Fundamental Intelligence?

Axioms of Evolution. As stated previously, there is little doubt that Evolution is a real process, and that Evolution has resulted in the development of the amazing variety of animal and plant species on Earth. Equally certain is the fact that Evolution is a result of at least two distinct processes; Intelligent Selection and Natural Selection. Further, it is clear that Intelligent Selection can provide faster and more responsive introduction, modification, or coadaptation of plant and animal species either as a matter of Intelligent Design, expedience, or even survival.

Natural Selection is cooperative with Intelligent Selection in the Evolutionary process, since it allows for the testing of the long term viability and ultimate success of those Species based on the strength of the characteristics.

In Summary, a complete study of Mr. Darwin's Theory on Evolution must recognize that two natural processes are at work in the development of all species and variants of living things. Intelligent Selection is clearly driven by an Intelligent Being through forces derived from Fundamental Intelligence. Natural Selection, driven by the Fundamental Interactions of Matter, is derived from a different Source? What might that Source be? What is the Trigger for Natural Selection? For this final question, we will turn back to Mr. Darwin for his thoughts as outlined in the concluding chapter of his great work.

> "To my mind it accords better with what we know of the laws impressed on matter by the Creator, that the production and extinction of the past and present inhabitants of the world should have been due to secondary causes, like those determining the birth and death of the individual."
> -Charles Darwin, *The Origin of Species*

Yes, Mr. Darwin credits the process of Natural Selection to the laws impressed on matter by the Creator. With this understanding of Evolution, the following Axioms are evident.

Axiom 15.1: Animal Species and Variation can result from what Darwin called Man's Selection, Artificial Selection, or Domestication; including the process now referred to as Selective Breeding

Axiom 15.2: Mules did not Evolve Through Natural Selection

Axiom 15.3: Many Animals did not Evolve Through Natural Selection

Based on the above Axioms, the following theories are offered for consideration.

Theory 15.1: Theory of Intelligent Selection: Intelligent Selection is one of two fundamental causes of the Evolutionary Process; and is an alternative to Natural Selection for the introduction of Species and Variation within the Plant and Animal Kingdom.

Theory 15.2: Theory of Intelligent Design: Intelligent Design is a process through which the decisions of an Intelligent Agent are used to implement Intelligent Selection.

CREATION

Every Life has a Beginning

What's in a Definition? The scientific study of the Creation of Life will be as controversial, if not more so, than the previous discussion on Evolution. Therefore, we will take the same approach and start with some basic principles and build on those.

Mr. Darwin's seminal work on the Theory of Natural Selection gave us a nice starting point for the discussion of Evolution. Unfortunately we have no such foundational scientific study on the Creation of Life. This might be because Restricted Science is generally reluctant to use the term Creation in the context of the Origin of Life. Therefore, we shall start with the definition of the term itself, Creation. In simple terms, Creation is the action, initiation, or process through which something comes into existence. This seems pretty easy to comprehend; but we can confirm the definition through a review of the steps in the Creation process.

Something "**X**" does NOT exist. An Event occurs, and as a direct result of the Event, Something "**X**" now exists; Something "**X**" has never existed before. In this case the new Something "**X**" came into existence and was clearly Created by the Event. This is definitional: It's what the term Creation means.

We shall therefore assume that in a scientific study of Creation, we can at least start with a common definition of Creation as stated above.

The next step in understanding Creation is to consider what CANNOT, as a matter of Physics, be Created. We shall look back to the work of Antoine Lavoisier to discuss the principle of Conservation of Mass. This is another Law of Nature that requires significant research for a full understanding; and we cannot cover the matter in detail herein. However, to paraphrase, the Principle of Mass/Matter Conservation states that in a closed system, the quantity of Mass within the system will remain constant. Combining this with our knowledge of the Mass-Energy Equivalence, the same can be said (and of course has been said) with regard to the Conservation of Energy. Since all Matter is made up of Mass, this boils down to the Principle that Matter cannot be Created or Destroyed, but only changed from one form to another. Hence, Matter, Mass, and Energy cannot be created! This leaves us with quite a problem, since nearly everything we know is Matter.

Or is it?

Forces exist, and they aren't Matter. Forces drive Behavior, and Behavior is not Matter. Is Intelligence Matter? More to the point of this discussion, is Life simply another form of Matter?

This is another fundamental question with profound implications. We know that all of the forms of Life on Earth occupy Matter. There is no question that each of us is comprised of atoms and molecular structures that define the Physical Form we see in the mirror. However, Life is not simply a physical formation of Matter. Life is the animation or Behavior of that Physical Form as it moves through SpaceTime.

To better understand the question and try to discern an answer, we shall observe two events that are at critical points in Life; **Death** and **Conception.**

Death

It seems odd to start a discussion on the analysis of Life with a study of Death. However, at that stage of Life, we get a clear delineation between Life and Matter.

All Life on Earth occupies a material form - Matter. We will refer to that matter herein as the Physical Form, as a distinction from the animated Behavior of the Form, which we call Life. This is an important distinction and warrants further clarification.

The Matter in which Life exists; its Physical Form, behaves in two completely different ways. When inhabited by Life Forces, the Physical Form demonstrates animated behavior, which we call Life. However, once Life Forces are removed the behavior of the Physical Form is completely different. It's the same matter, yet its behavior is that of a completely inanimate object. At some point, every Life will leave, or cease to occupy, every Physical Form. Yet in every such instance, the Physical Form remains intact immediately following the loss of Life (unless transformed by some outside agent).

Based on the Conservation of Mass, the Physical Form, composed of Matter, is neither created nor destroyed. It will simply be transformed to some other Mass. However, the Life is absolutely gone from the Physical Form. The Forces that drove that Life, whether those Forces resulted from Fundamental Interactions or Fundamental Intelligence, will never return.

It is important to recognize that the Fundamental Interactions that hold the matter of the Physical Form together continue to exist after the Force of Life has departed. The atoms and molecules of the matter remain intact. Therefore, there must be a separation between the Force of Life and the Forces of Physical Form - Matter. Otherwise the entire Physical Form would fly apart when Life ceased. The matter would disappear as if hit by a magic wand.

Equally important, based on the Conservation of Energy, the Mass and Energy of the Physical Form are equal immediately before and immediately after the departure of Life. No Energy or Mass was destroyed. Yet something clearly changed. Cellular structures continue to exist; but are no longer driven by the Force that causes replication. The engine that converted the Energy of the Physical Form into animated behavior is gone.

There is another important observation when it comes to Death. As soon as Life departs from the Physical Form, Fundamental Interactions of Matter begin to take over to transform the matter. Fundamental Interactions start to drive the deterioration of the Physical Form. These Forces can be chemical or biological, and could involve interaction with other life forms as the Matter decomposes. What's important to note here is that Life strongly resists the natural decomposition of the Physical Form as long as Life occupies the Physical Form. If Life is removed from the Physical Form, or even cut off from a portion of the Physical Form, the resistance to this decomposition of Matter is completely removed.

This illustration demonstrates that Life, and particularly the Forces that drive Life, are not subject to the limitations of the Conservation of Mass, Matter, or Energy. Life cannot be solely comprised of Matter, Mass, or Energy; or the Fundamental Interactions. Life is something that inhabits the Physical Form; but is not the Physical Form.

Clearly, the matter in all of the Physical Forms of Life behaves very differently when Life is present than it does as soon as Life is removed from that Physical Form. Therefore, Life Forces are independent of, and can actually work to resist the Fundamental Interactions that drive Matter. We may therefore conclude that Life and Life Forces are not governed by the Conservation of Mass/Matter, and therefore may be subject to Creation.

Conception

For our next observation, we will look at the other end of Life on earth, Conception. This observation does not include single cell replication, single cell reproduction, or other forms or reproduction that do not result from the process of fertilization.

Conception involves combining two Gametes to produce a new Zygote, the first cell in the earliest stage of Life. Neither Gamete is an organism. They do not contain the full set of chromosomes that a Somatic Cell does. However, as soon as the two Gametes join, Life is initiated, a Zygote is produced.

The zygote is certainly not a Living Being. However, this single cell has all the elements necessary to define the Life that will emerge at some point from the biological process that will drive it forward. The zygote will begin splitting into additional cells, and the Physical Form that contains Life starts to grow.

Every Life is New. Unlike all of the other Mass/Matter of the Universe, which existed at the Big Bang and which is only subject to a change or transformation of other Matter; the fusion of the two gametes Creates something that has never existed before. The genes and chromosomes of the new Life are unique and individual.

There will be those that argue that this new Physical Form is in fact merely a new arrangement of existing Matter; that all Life is comprised of the same substances, just arranged differently. This argument collapses based on the existence of a single molecule; Deoxyribonucleic Acid - DNA. DNA has unlocked many of the secrets of Life. It is the "blue print" for the Individual Life that occupies that Physical Form.

DNA is a complex molecule. In fact, DNA most often exists as a pair of complex molecules intertwined in an elegant Double Helix structure. However, DNA's secret lies not in its complexity, but in its individuality; its singularity. Unlike all other matter, each Living thing is driven by a unique, exclusive DNA molecular structure. Even in Monozygotic Twins, recent studies demonstrate a Structural Variation in the nearly identical DNA affecting Copy-Number-Variation Profiles. This confirms that DNA always exists as a unique molecular structure for every living thing.

This is another absolutely phenomenal observation that must be considered carefully. In the trillions of trillions of Living things that have occupied Earth, each living thing has a distinct, unique, individual molecule that was produced solely for them. To even further add to the phenomenal nature and complexity of DNA, each New DNA Molecule for every New Living Thing is Created. Yes, by definition, Created.

There are two processes for production of the DNA Molecule. The first is **Replication**; the second is **Creation**.

DNA Replication. DNA Replication is the easiest to understand. It occurs in every living multi-cellular organism as part of the replication of Somatic Cells through a life-cycle cell division process know as Mitosis. In Mitosis, the twin DNA molecules comfortably intertwined in their double helix structure within a cell begin to separate, or "unzip".

This creates two tines similar to a tuning fork. As they unzip, enzymes cause bases to start to attach to the "unzippered" tines, or strings. Each separating string becomes the structure by which a new duplicate DNA molecule is produced. The replication process produces two identical DNA double molecules, allowing for cellular division to create two identical Somatic Cells.

DNA Creation. DNA Creation is a bit more complicated, and actually requires two individual and isolated steps that are separated by space and time. The two steps in DNA Creation are Gamete Production and Fertilization.

We will start with the first step, production of a Gamete through Meiosis. Meiosis produces a Gamete through a 15 step process; versus the 8 steps of Mitosis. Instead of producing two duplicate somatic cells, Meiosis produces specialized Haploid Cells containing one half the chromosomes of the parent. Females produce egg cells and males produce sperm cells.

DNA undergoes an extraordinary transformation in Meiosis; a process called Recombination; including Chromosomal Crossover. Without going into detail, the DNA molecule breaks and rejoins into strands that start the process of creating new DNA. Recombination provides each of the new Gamete Cells with a unique half-set of chromosomes that contain the building blocks for a new and unique Life. Once Meiosis is completed, new Gametes are stored in the parent, standing by for the next step in the process, Fertilization.

Gametes are not individual living organisms. They are specialized cells of the parent, contained in the parent. Gametes do not include a full set of chromosomes, nor do they contain a complete, unique DNA Double Helix. New Life depends on the fusion of two Gametes through the process of fertilization.

When two gametes combine through Fertilization, the initial cell of a new Life is formed - a Zygote. Much of the matter in a zygote is the same as the matter in every other zygote; proteins, enzymes, membrane structures, and other ingredients. However, that matter does not define Life, or the behavior of the Living thing. Instead, the fusion of the gametes creates a brand new molecule that has never existed before - New DNA. There will be those who question the use of the word "Creates", so I will restate the process of Creation and see if the two comport.

In the act of Creating, an Event occurs, and as a direct result of that Event, something that has never existed comes into existence. So, as you can see, by definition, the DNA for every new Living Thing is Created for that Living Thing. It has never existed before.

Of equal importance is the fact that the DNA Creation process is neither random nor accidental. The processes of Meiosis, Recombination, and Fertilization are highly complex and exceptionally accurate. The molecular events of Recombination lead to elegant rearrangements of DNA strands that do not appear to flow from the random mutations of Natural Selection.

Recombination produces functionally perfect DNA strand variants at the exact same time, in the exact same parent. Also, while the specific DNA molecule of every living thing is unique, the structure is precise and the construct follows a pattern directly from the contributing parents. We are able to trace each DNA molecule back to parents, and even identify closely aligned offspring; or hereditary branches. The entire process of New DNA Creation demonstrates a highly ordered, accurate, and repeatable, series of actions that define an Intelligent Behavior.

Life is Created

As stated before, Charles Darwin, without knowing anything about DNA, observed intelligent interaction with the process of New DNA Creation in the Domestication of Living Species. This is the process he referred to as Man's Selection and which we have defined as Intelligent Selection. Scientific discoveries over the last century have clearly demonstrated that both the Behavior and Physical Form of every Living Thing are to a large part driven by the pair of unique and individual DNA Molecules; DNA that was produced through an intricate, detailed, and highly predictable process specifically for the Life it governs.

Under Intelligent Selection, Human Intelligence utilizes this highly defined New DNA Creation process to Design and Create New DNA that in concert with the rest of the Matter in a Zygote, results in a New Living Thing. Darwin recognized this in the long history of Man's domestication of animals. The New Living Thing could be a refinement of the Parent, such as a Race Horse; a variant of the Parent, such as a Dog Breed; a cross hybrid Species of two different Species, a Mule; or an entire new Species within a larger Family that thrives and allows for its own set of refinements, variants, and hybrids, such as Cattle. Each of the above examples demonstrates the Force of Intelligent Design in the Creation of a New DNA Molecule, and each New DNA Molecule results in the Creation of a New Living Thing.

Could the same Intelligent Force that is clearly evident in the domestication of plants and animals under Human Intelligence be at work in Nature. Once again, we will look back to the observation of the Father of Evolution, Charles Darwin.

> "Can the principle of selection, which we have seen is
> so potent in the hands of man, apply under nature? I
> think we shall see that it can act most efficiently."
> - Charles Darwin, *The Origin of Species*.

Clearly, Mr. Darwin thought so.

The only question remaining is whether Intelligence can exist in the Universe outside of Human Intelligence, or the intelligence of other living things on earth. If the answer to this is yes, then it is clear that according to Darwin, the Intelligence found elsewhere in Nature could act efficiently in the exercise of Intelligent Selection in the Creation of New DNA; in the Creation of New Living Things.

Axioms of Creation. Coupling Scientific Knowledge of DNA with the above definition of Creation, the following Axioms are evident.

> **Axiom 15.4: Human Intelligence can use Intelligent Selection to Design and Create Living Things that meet specifically desired objectives of Behavior and/or Physical Form.**

> **Axiom 15.5: Intelligent Selection can Create Living Things that are new Species or variants of existing Species.**

Based on the above Axioms, the following Theory is offered:

> **Theory 15.3: Theory of Intelligent Creation: Living Things, both Plant and Animal, can be Created through the application of Intelligent Design through the process of Intelligent Selection.**

Based on the additional Axioms and Theories derived from an understanding of Evolution and Creation, the Postulate of the G-Phenom is refined as follows:

> **Postulate: The G-Phenom is the Singular Supreme Fundamental Intelligent Being that Initiated the Creation of the Universe and Triggered the Big Bang; and is the Source of Intelligent Forces within the Universe, including the Intelligent Forces driving the Creation and Evolution of Life.**

Chapter 16 - Life Equilibrium: A Happy Place

LIFE FORCES

The Power of Intelligence. We have come a long way in this study, but it all brings us back to our original Questions. Why are we here? What causes Life to exist? How did it all start?

By now, it has become very clear that nothing can happen over time without Forces. Matter cannot exist without Forces. Bridges and buildings could not stand up without Forces. There would be no Behavior without Forces. Life could not exist without Forces. It is apparent that Balanced Knowledge of Forces is pretty important. Equally important is our willingness to continuously Learn about Forces and how Forces drive Behavior.

Current Restricted Science derives all of Nature's Forces from the Fundamental Interactions of Matter; Gravity, Strong, Weak, and Electromagnetic. However, limiting the initiation of Forces to the four Fundamental Interactions of Matter leaves large gaps in our understanding of observed Behaviors in both the Universe and Life. New Theories of Matter, Mass, and Energy have been introduced in an attempt to plug these gaps. However, there has been little discussion on new Theories of Forces, or consideration of other Fundamental Phenomenon that would initiate Forces that could in fact provide more reasonable and elegant answers to the observed Behaviors.

The conflict between observed Behavior and Fundamental Interactions is exacerbated when the Phenomenon of Life is introduced to the Universe. Living Things demonstrate Behaviors that simply cannot be traced back to the Forces of Fundamental Interactions. More importantly, Living Things appear to be able to make independent choices on Behavior; choices that flow from the

Force of Intelligence. The observation of these Behaviors provides strong evidence in support of the Theory of Intelligent Forces.

Another equally compelling evidence of the Theory of Intelligent Forces flows from observations regarding Equilibrium. Human Behavior is one of the most dynamic Behaviors observed in our Universe. Human Interactions can result in sudden and volatile fluctuations in the oscillations of Human Behavior, rapidly upsetting the Equilibrium of the systems. In large populations of Humans, these systems can be political, social, and economic. In individuals the sudden changes in Equilibrium can impact physical, emotional, or psychological systems.

The more we understand our Universe and Life, and the Forces that drive Human Behavior, the better equipped we are to employ Intelligence to find Equilibrium in Human Interactions; the Happy Place. We will significantly improve our ability to address these complex issues if we recognize the existence of Fundamental Intelligence, as well as the Source of Fundamental Intelligence; the Supreme Intelligence of the G-Phenom.

This was clearly recognized by our Founding Fathers in the Declaration of Independence, which validated the importance of the "Laws of Nature and Nature's God". Ironically, our current refusal to allow the discussion of these Theories as part of science leaves the Force of Fundamental Intelligence open for the exploitation by godless social science experiments of for abuse by highly restricted and fanatical Theocracies. Humans have clearly demonstrated that they can use the same awesome power of Intelligence to upset Equilibrium in order to strengthen their own perceived power or authority.

It is essential that we study the power of Fundamental Intelligence to determine the best approach to the design of Human Interaction systems to allow for the oscillations that are required to maintain stability, but dampen the extreme interactions of fringe Behaviors that are intended to upset Equilibrium.

THE CHOICE: KNOWLEDGE OR IGNORANCE

This is the single most important reason to consider the Theory of Supreme Intelligence and the Theory of Intelligent Forces as a matter for scientific study. If we deny the Forces that drive Intelligent Behavior, we have no hope of designing effective systems for dealing with that Behavior. Most Humans would like to achieve Equilibrium - the Happy Place. However, there are those who would prefer to disrupt the equilibrium in the hope that they would gain an advantage in controlling the Power that flows from Intelligence (remember Hitler?).

In the most extreme cases, we observe fanatics who employ Intelligence to wreak havoc through terrorism. These horrendous Behaviors are not a result of Gravity or Electromagnetic Interactions. They are certainly not the inevitable or preordained Behavior of a Matter based, Deterministic Universe. Instead, they are the result of an atrocious misuse of the Forces of Intelligence by wicked people. The entire goal of these Behaviors is to cause such extreme volatility in Equilibrium as to completely change a system.

If, as a matter of public scientific study, we choose to remain Ignorant regarding the force of Fundamental Intelligence, the Knowledge of the Power of these Forces will be left in the control of Restricted Tradition, Restricted Reason, or Restricted Experimentation; all of which have resulted in disastrous consequences in past Human Interactions.

ACHIEVING EQUILIBRIUM

Life is driven by forces. The Universe is driven by forces. When these forces are balanced, we achieve a state of Equilibrium that allows for the improvement in the Human Condition, and a greater Stewardship of Planet Earth. All of the Forces flow from naturally occurring phenomena that comprise Nature's Laws. There is clear and compelling evidence supporting the Theory of Intelligent Forces and the Postulate of the G-Phenom as the Supreme Intelligent Being from which Fundamental Intelligence flows. Fundamental

Intelligence produces the Forces driving Intelligent Behavior; Forces that cannot be explained through the Fundamental Interactions of Matter.

If we are to achieve and maintain Equilibrium in Human Interaction, and find answers addressing the challenges facing the Human Condition, we must consider scientifically the Forces of Fundamental Intelligence; and the Source of all Fundamental Intelligence; a Supreme Intelligent Being, the G-Phenom.

Chapter 17 - Communicating Intelligent Action

How?

For the sake of the argument in this Chapter, we shall assume that Fundamental Intelligence is accepted as a possibility and the G-Phenom is a viable solution to the Theory of Everything. Therefore, there is a Singular Fundamental Intelligent Being at work in the Universe. As both a Singularity and the Source for all Intelligence, the G-Phenom would be the Supreme source of Intelligence for our Universe; the Supreme Intelligent Being. How can this Intelligent Being communicate with other Intelligent Beings throughout the Universe? Further, how can this Intelligent Being possibly effect intelligent action on other matter or Beings in the Universe.

These are certainly challenging questions; and both philosophical Reason and religious Tradition can be instructive in this subject. However, this is a scientific study, and therefore we will use engineering principles alone to shed some light on this important topic. We will examine two specific behaviors that are observed as flowing from Intelligence; the desire to communicate Intelligence ("Communication") to other Intelligent Beings, and the desire to effect actions or cause Behaviors ("Action") as a result of Intelligence.

Communication. This is the easier of the two behaviors to evaluate. Communication between Intelligent Beings crosses numerous channels and uses a variety of senses.

Humans' most used (and often misused) communication channel is speech. We can talk, and sometimes we can even listen. Sight is another very effective communication channel. Humans can read; but they can also pick up nuances of communication in a glance; or through a wink of an eye. These are very popular communication channels. However, when compared to many other animals, Human communication channels are often much less powerful or effective.

Dogs can communicate through smell, using olfactory tags to communicate territorial boundaries. Dolphins communicate using a highly specialized sonar capability to hunt for food, avoid predators, and play. Horses are highly sensitive to the slightest vibrations in their environment, and can "sense" approaching danger well before other senses detect the threat. Animals can detect changes in weather long before Humans can. Highly specialized communication channels have developed throughout the animal kingdom for the safety, security, survival, and even pleasure of the Species.

Man has also used the electromagnetic spectrum to communicate in ways that, only a few centuries ago, were unimaginable. Intelligent Beings are in constant communication with other Intelligent Beings over very broad distances and across all spatial dimensions. As we begin to understand the SpaceTime Continuum, the possibility of communicating across the temporal dimension becomes more intriguing. Think of the ramifications of opening this communication channel. It may be possible at some point in time to confer with Mr. Einstein or Mr. Newton - maybe even a three-way conversation with both men.

Human Intelligence is just starting to scratch the surface of the concept of Communication. We have so much to learn from our own Intelligence alone, not to mention what we might gain from the communication advancements of other forms of Life on Earth. As we open our minds to the capacity of a Supreme Fundamental Intelligence, we can see that the possibilities become limitless.

Action. As the old saying goes, Actions speak louder than Words. While Communication is important, if we are unable to Act upon the Communication, it is simply information. If Intelligent Beings can communicate across the entire SpaceTime Continuum, can we execute actions based on that Communication? In other words, can we make things happen only through Communication?

Of course, the answer is yes. We do it every day with our television remote control or a distant space vehicle. Intelligent Beings send information to an Intelligent Agent (the TV or space craft) that executes an Action or exhibits a Behavior as a direct result of the

Communication. But can we make a similar connection at a Fundamental Level, without the aid of physical constructs or machines. In other words can an Intelligent Being, Supreme or otherwise, communicate an action or behavior with another Intelligent Being directly.

Recent scientific study indicates that indeed such communication can occur. The Theory behind this level of communication is called **Quantum Entanglement.**

Quantum Entanglement

The Theory of Quantum Entanglement is neither superstition nor the Boogeyman. It is a Scientific Theory and study that offers new understanding of the way things behave.

This Theory would allow that if Intelligent Beings derived their Intelligence from the same source; say a Supreme Fundamental Intelligent Being; then even after these multiple Intelligences are separated, they can remain "Entangled" with regard to certain properties or behaviors. This is clearly an oversimplification of a complex subject; however, it illustrates the point.

We can examine a hypothetical possibility of Quantum Entanglement at work in Intelligent Selection. There is no question that Intelligent Selection can Create specifically Designed Life Forms. Again, Mr. Darwin made this point clear. What if, through Quantum Entanglement, a Supreme Fundamental Intelligent Being introduced a new Intelligent Being through the Creation of New DNA in a Zygote? Could the G-Phenom use Quantum Entanglement to Create, through Intelligent Selection, a new Life Form?

The more important question for this Study is why couldn't that happen? We know that Man can Create new Life forms through Intelligent Selection. We believe that Quantum Entanglement is a real Scientific Phenomenon. The Theory of Fundamental Intelligence is supported by compelling evidence. The Postulate of the G-Phenom is a very real possibility. Why couldn't a Supreme Fundamental Intelligence interact with Intelligent Beings through Quantum Entanglement? Why would we not accept the possibility that Nature's Laws are subject to Interaction with Nature's God?

This topic alone is exhaustive, and relies on all three Learning Channels for Balanced Knowledge. We must consider not only our scientific Experimentation, but also our Reason and Tradition to further our understanding in this area. My goal for this short chapter was not to convince you of the interaction between God and other Intelligent Beings through Quantum Entanglement. Instead, my sole purpose was to whet the appetite for the continued scientific study of God, and open our minds to the possibilities that such a Study would allow.

PART SIX: From Infinity to Beyond

Chapter 18 - From Infinity

INFINITY

The Limits of Infinity. The concept of Infinity is almost impossible to comprehend. Yet this has never stopped Human Intelligence from wondering about Infinity. We stare into the heavens and wonder where it ends. We think about time, and wonder where it begins. We try to see as far as we can, and think about what's beyond.

Through the application of Intelligence, we have made significant progress in tackling this almost unfathomable concept. For example, mathematics employs the concept of Infinity in a very practical way - through calculus. We integrate the area bounded by a curve by taking the summation of infinitesimally small sections under that curve. Only by approaching the Limits of Infinity can we solve the equation accurately.

We can also see the practical approach to Infinity in analyzing asymptotic Behavior. This Behavior considers a continuous approach to a Limit without ever arriving at the destination. With each step in time, the movement takes another step closer to the Limit; yet it will never get there. As you can see, each step becomes smaller and smaller, since the distance to the limit gets smaller.

The above examples demonstrate that while the concept of Infinity seems almost incomprehensible, Humans have, through the Force of Intelligence, harnessed the Limits of Infinity for practical applications. These two concepts deal with spatial Infinity. Over the last century, Human Intelligence has started to tackle concepts of temporal Infinity. Concepts such as Time Dilation allow for temporal dimensional variation; an initial step in better understanding this dimension. Future discovery may offer the possibility of multiple temporal dimensions. Could Time have both a length and a width?

Infinity is truly mind-boggling. However, Intelligence allows us to not only consider and reflect on Infinity, but also to harness the power of the concept of Infinity for practical applications.

SUPERIOR INTELLIGENCE

One of the strongest arguments in support of the Theory of Supreme Intelligence and the Postulate of the G-Phenom is the possibility that there exists an Intelligent Being in the Universe that possesses greater Intelligence than Man. The issue has been addressed several times in this Study, and the evidence of such Superior Intelligence in both the Origin of the Universe and the Creation of Life is compelling.

For some reason, however, many in the Restricted Scientific community will simply not accept that such a Superior Intelligence can exist. These same Scientists ponder Intelligent Life elsewhere in the Universe with little reservation. They also accept the possibility of multiple unobserved dimensions in SpaceTime. Unfortunately, as soon as the concept of a Supreme Intelligent Being or Superior Intelligence is introduced; the scientific discourse shuts down. Instead, Restricted Science continues to focus on the Fundamental Laws of Matter for solutions to the most challenging questions posed by Intelligent Life.

Why are we here? How are we to behave with respect to the Earth and fellow man? How can we achieve Equilibrium in Human Interactions? Does the Chiropractic concept of Innate Intelligence offer healing forces that we might be able to unlock if we simply consider them?

These are tough questions demanding the practical application of Intelligent Solutions. Unfortunately, Restricted Science is looking for answers to these pressing challenges solely through inanimate sub atomic particles and arcane string theories.

The combination of Science, Philosophy, and Tradition has allowed Human Intelligence to approach the concept of Infinity and develop practical applications using that concept. For example, much of Newton's work in the advancement of Mathematics was based on Natural Philosophy and Tradition. Nevertheless, scientific study in the last century has all but ruled out Philosophical and Traditional concepts of Superior Intelligence as a Fundamental Force in Nature.

The power of Human Intelligence is evident, both in the good and evil of Human Behavior. If we allow the Study of Fundamental Intelligence to stay in scientific darkness, then the Power of Fundamental Intelligence will be relegated to those who would abuse it. If, however, we open our minds to the possibility of a Superior Intelligence, then we can expose that Power to the Light of Scientific Study; and quite possibly find answers to the challenges of Human Interaction and advance the Human Condition.

INFINITE INTELLIGENCE

Human Intelligence is truly a phenomenon of Human Life. But where can Human Intelligence take us? What are the limits of the Power of Intelligent Forces?

If Intelligence is merely a result of the fixed and inanimate Forces derived from Fundamental Interactions working in the grey matter between our ears; then Intelligence must be finite. At some point we will use up the power of our Matter Computer; our brain. If however we accept the concept of Fundamental Intelligence as a separate Universal Phenomenon, how much Intelligence is available to tap? How far can we push ourselves to answer the Questions that perplex us?

This Engineering Treatise has attempted to demonstrate a plausible theory with compelling evidence that the gift of Human Intelligence flows from Fundamental Intelligence, the Source of which is a Singular Supreme Fundamental Intelligent Being; the G-Phenom. This would beg the question, how Intelligent is Supreme Intelligence? How deep is the well from which we can draw intelligent water? As with the scientific study of the Beginning of the Universe, we will extrapolate backward in Time to try to get a sense of the depth of Fundamental Intelligence.

Human Intelligence has inhabited Earth for about 500,000 years.

The G-Phenom has existed since at least the Beginning of the Universe, about 13.75 Billion years. We shall refer to that time frame as the Universal Age. Therefore, Human Intelligence has been developing for only about 36 millionths (0.000036) of the Universal Age. That equates to much less than a blink of an eye.

We will assume that the original Humans had sufficient Intelligence to survive; otherwise they wouldn't have. So, that will serve as our base line for Human Intelligence. Humans have therefore advanced their Intelligence from the baseline of minimal survival skills to the current level of understanding in 0.0036% of the Universal Age. We might also observe that the growth in Human Intelligence is an accelerating curve. The more Intelligence we gain, the faster we gain Intelligence.

So, we can get an understanding of the potential for growth of Human Intelligence by plotting the actual growth in Human Intelligence over the entire Universal Age. Let me warn you; it will take a really big piece of graph paper! Suffice it to say, we begin to approach the Limits of Infinity again.

The Supreme Intelligence of the G-Phenom is Infinite. The answers to the Question posed at the beginning of this book are contained within the Theory of Supreme Intelligence. We cannot focus on our own limited knowledge for the answers – our knowledge is clearly fallible. We must continually challenge ourselves to open our minds and our Learning to the Supreme Fundamental Intelligence of the G-Phenom

If we are to tap the full potential of Human Intelligence, it is essential that we recognize that our Intelligence is not Restricted to the Fundamental Interactions of Matter; that the answers to many of the challenges we face are not based on what Science can do with Matter; but what Intelligence can do with Behavior. We must not restrict the science of Creation to the behavior of a Higgs Boson. We must make room for the scientific study of the Theory of Supreme Intelligence and the Postulate of the G-Phenom.

Chapter 19 - To Beyond

BEYOND THIS BOOK

The purpose of this book was not to answer questions. It was to spark discussion. The Theories, Axioms, and Postulates presented herein have been supported with scientific analysis. However, they are just a start. If this book is to have an impact then we must use this as the beginning of a revitalized public discussion on the Origin of the Universe and the Creation of Life.

AMERICA'S FOUNDATION

The United States was founded on the principle that certain unalienable rights flow from a Higher Authority - a Supreme Intelligent Being. At the same time, our Founding Fathers recognized the need for tolerance for all individual beliefs. Each of us is entitled to pursue any faith tradition we choose, or no faith tradition at all. These two concepts are not mutually exclusive. God existed before Religion, and God exists in the absence of Religion.

Current Restricted Scientific Study however has actually taken a proactive stance against the existence of a Supreme Fundamental Intelligent Being, even in the light of clear and compelling evidence supporting such a Theory. We have allowed the religion of Atheism to dominate the last Century of scientific study.

The U.S. Declaration of Independence recognized the Unalienable Human Rights endowed by the "Creator". The U.S. Constitution guarantees the "Blessings of Liberty" as the basis for Human Interaction and the Rule of Law. It is therefore impossible for our Nation to serve as the Beacon of Liberty and the Champion of the Rule of Law without recognizing the existence of God; without acknowledging that Liberty is a Blessing. How is it, that in classrooms across the Nation we can study the fundamental of Human Rights and Nature's Law, but cannot even discuss the Creator or Nature's God?

If the United States is to remain, in the words of Ronald Reagan, the Shining City on the Hill, we must maintain the strong Foundation for both the Authority of Nature's Laws and Tolerance of the Blessings of Liberty that our Founding Fathers established. We must recognize that Unalienable Rights are not granted by man: They are Endowed by the Creator.

GOD IS: A MATTER OF FACT

God is not a matter of fantasy or superstition. God is not a philosophical construct or religious legend. God is certainly not a figment of imagination. GOD IS A Matter of Fact, and therefore a subject for scientific study!

Fundamental Intelligence is the most powerful phenomenon in the Universe. The Scientific Evidence is both clear and compelling. The Postulate of the G-Phenom defines a Supreme Fundamental Intelligent Being as the Source of Fundamental Intelligence; the Force that Triggered the Big Bang, Originated the Fundamental Interactions of Matter, and Created Life. In the United States, we acknowledge this Being in our National Motto: "In God We Trust". You can call this Supreme Fundamental Intelligent Being anything you want to. By any other name, Fundamental Intelligence is real; the G-Phenom exists. GOD IS.

Fundamental Intelligence plays a vital role in the Behavior of the Universe, the evolution of Life, and the complexity of Human Interaction. Humans have been entrusted with the highest level of Intelligence on Earth; an amazing privilege as well as an awesome responsibility. Man has used this gift in wonderful ways to advance the Human Condition. Man has also used this gift to inflict horrific pain and suffering.

Our choice is clear. We can continue to allow Restricted Science to prevent the introduction of such Theories as Fundamental

Intelligence and the G-Phenom, and leave the power and authority of God in the hands of limited and often fanatic Theocracies. Or we can open scientific study to the possibility that a Singular Supreme Fundamental Intelligent Being does have a hand in the affairs of the Universe and the Human Condition.

START THE DISCUSSION NOW

The Theories and Postulates presented herein are Science. This Engineering Treatise and the concepts discussed provide the basis for compelling scientific study in every high school in the United States, and any other appropriate forum. We must freely and openly discuss God as a Phenomenon of the Universe; a Matter of Fact.

Only then will we be able to answer questions regarding Human Behavior; seek solutions to the challenges in Human Interaction; and advance the cause of the Human Condition and Planet Earth.

GOD HELP US IF WE DON'T!

PART SEVEN: Appendices

Axioms, Theories, and Postulates

AXIOMS PRESENTED

Axiom 4.1: Experience is what we are exposed to over time.
Axiom 4.2: Humans only Learn through Experience

Axiom 6.1: Everything is Something unless it is Nothing.
Axiom 6.2: Something cannot come from Nothing

Axiom 7.1: Everything came from Something
Axiom 7.2: The Universe was Created by Something
Axiom 7.3: The Big Bang was Triggered by Something

Axiom 11.1: Forces are not Self Initiating
Axiom 11.2: Interactions initiate Nearly All Forces
Axiom 11.3: Forces can be Initiated by the Intelligence of
    Intelligent Beings

Axiom 15.1: Animal Species and Variation can result from
    what Darwin called Man's Selection, Artificial Selection,
    or Domestication, including the process now referred to
    as Selective Breeding.
Axiom 15.2: Mules did not Evolve Through Natural
    Selection
Axiom 15.3: Many Animals did not Evolve Through Natural
    Selection
Axiom 15.4: Human Intelligence can use Intelligent Selection
    to Design and Create Living Things that meet specifically
    desired objectives of Behavior and/or Physical Form
Axiom 15.5: Intelligent Selection can Create Living Things
    that are new Species or variants of existing Species

THEORIES PROPOSED

Theory 1.1: The Theory of Supreme Intelligence: A Singular Supreme Intelligent Phenomenon Initiated the Universe and Originated Life.

Theory 11.1: Theory of Intelligent Forces: Forces can be Initiated by the Intelligence of Intelligent Beings.

Theory 15.1: Theory of Intelligent Selection: Intelligent Selection is one of two fundamental causes of the Evolutionary Process; and is an alternative to Natural Selection for the introduction of Species and Variation within the Plant and Animal Kingdom.

Theory 15.2: Theory of Intelligent Design: Intelligent Design is a process through which the decisions of an Intelligent Agent are used to implement Intelligent Selection.

Theory 15.3: Theory of Intelligent Creation: Living Things, both Plant and Animal, can be Created through the application of Intelligent Design through the process of Intelligent Selection

THE POSTULATE OF THE G-PHENOM

The G-Phenom is the Singular Supreme Fundamental Intelligent Being that Initiated the Creation of the Universe and Triggered the Big Bang; and is the Source of Intelligent Forces within the Universe, including the Intelligent Forces driving the Creation and Evolution of Life.

Index